MATHEMATICS

THE NEW YORK UNIVERSITY LIBRARY OF SCIENCE

MATHEMATICS

EDITED BY

Samuel Rapport AND *Helen Wright*

ACADEMIC EDITORIAL ADVISER

HOLLIS R. COOLEY

Professor of Mathematics, Washington Square College,
New York University

NEW YORK UNIVERSITY PRESS 1963

ACKNOWLEDGMENTS

"Archimedes" from *A History of Science* by George Sarton. Copyright 1959 by the President and Fellows of Harvard College. Reprinted by permission of Harvard University Press.

"Newton" from *Makers of Mathematics* by Alfred Hooper. Copyright 1948 by Random House, Inc. Reprinted by permission of Random House, Inc. and Faber and Faber Ltd.

"The Prince of Mathematicians: Gauss" from *Men of Mathematics* by Eric Temple Bell. Copyright, 1937, by E. T. Bell. Reprinted by permission of Simon and Schuster, Inc.

"Evariste Galois" from *The Life of Science* by George Sarton. Copyright 1948 by Henry Schuman. Reprinted by permission of Abelard-Schuman Limited.

"Einstein's Great Idea" by James R. Newman, originally published in *The Saturday Evening Post*. Copyright © 1959 by The Curtis Publishing Company. Reprinted by permission of the author.

"William Rowan Hamilton" by Sir Edmund Whittaker, originally published in *Scientific American*, May, 1954. Copyright © 1954 by Scientific American, Inc. and reprinted by their permission. All rights reserved.

"The Nature of Mathematics" from *An Introduction to Mathematics* by A. N. Whitehead. Copyright 1948 by Oxford University Press, Inc. A Galaxy Book. Reprinted by permission.

"A Mathematician's Apology" from *A Mathematician's Apology* by G. H. Hardy. Reprinted by permission of Cambridge University Press.

"Mathematical Discovery" by Henri Poincaré from *Science and Method*, translated by Francis Maitland.

"Big Numbers" from *One, Two, Three . . . Infinity* by George Gamow. Copyright 1947, © 1961 by George Gamow. Reprinted by permission of The Viking Press, Inc. and Macmillan & Co. Ltd.

"Symbols" by Tobias Dantzig. Reprinted with permission of The Macmillan Company and of George Allen & Unwin Ltd., from *Number* by Tobias Dantzig. Copyright 1930, 1933, 1939, 1954 by The Macmillan Company.

"Geometries" from *The Anatomy of Science* by Gilbert N. Lewis. Copyright 1926 by Yale University Press, and reprinted by permission of the Press.

"Change and Changeability—the Calculus" from *Mathematics and the Imagination* by Edward Kasner and James Newman. Copyright 1940 by Edward Kasner and James Newman. Reprinted by permission of Simon and Schuster, Inc. and G. Bell & Sons, Ltd.

"Topology" from *What is Mathematics* by Richard Courant and Herbert Robbins. Copyright 1941 by Richard Courant. Reprinted by permission of Richard Courant.

"Mathematics and Modern Civilization" from *Introduction to Mathematics* by Cooley, Gans, Kline and Wahlert. Copyright 1937, 1949, by Hollis R. Cooley, David Gans, Morris Kline and Howard Wahlert. Reprinted by permission of and arrangement with Houghton Mifflin Company, the authorized publishers.

"From Cyzicus to Neptune" from *The Handmaiden of the Sciences* by Eric Temple Bell. Copyright 1937 by the Williams & Wilkins Company, and reprinted by their permission.

"The Geometry of the Epeira's Web" from *The Life of the Spider* by Henri Fabre. Copyright 1912 by Dodd, Mead & Company, and reprinted by their permission.

"Why Kitty Lands Butter-Side Up" by E. E. Slosson, from *Keeping Up With Science* by E. E. Slosson. Copyright 1924 by Harcourt Brace & Co. Reprinted by permission of Harcourt, Brace & World, Inc.

"The Laws of Chance" from *Facts from Figures* by M.J. Moroney, reprinted by permission of Penguin Books Ltd.

"Computing and Other Machines" from *Minds and Machines* by Wladyslaw Sluckin, reprinted by permission of Penguin Books Ltd.

"E = MC²" by Albert Einstein from *Source Book in Astronomy 1900–1950*, edited by Harlow Shapley. Copyright 1960 by the President and Fellows of Harvard College. Reprinted by permission of Harvard University Press.

CONTENTS

IV. Mathematics and the World Around Us

FOREWORD

MATHEMATICS is loved by many, disliked by a few, admired and respected by all. Because of their immense power and reliability, mathematical methods inspire confidence in persons who comprehend them and awe in those who do not. It is natural that every intelligent person should want to know something about this great field, and about the scholars who developed it. Many individuals who acquired a distaste for mathematics and lost confidence in their ability to understand it, because of inadequate teaching or other misadventure in their early schooling, discover a great desire to fill this gap when they have become more mature. This is most desirable, and it is an excellent thing to make available to them, as well as to others, such selections from mathematical writings as are here presented.

It is a commonplace today that science is continually becoming more mathematical. Physics is already mathematized to a high degree. In the other physical sciences, mathematical methods are being utilized more and more. The social sciences, too, tend in this direction. A thorough grounding in mathematics may soon be an essential prerequisite for serious work in any of these fields. Moreover, intelligent decisions in many areas of life must be based on knowledge which is mathematical in character, and this will become progressively more apparent in the future. It is most important that the thinking of those who develop this knowledge be understood and appreciated as much as possible by the general public which, in a democracy, is called upon to make the ultimate decisions. It is important that the layman have confidence in the scientists and economists, and not distrust them. To this end he must try to comprehend the method of thought which they employ, which is increasingly the mathematical method. The present book contributes to this purpose.

So short a book of readings as the present one can give only a glimpse, here and there, of the world of mathematics. But it will introduce the reader to ideas which he will want to pursue further. A few of the excerpts will acquaint him with the thinking of mathematical masters. Others describe bodies of mathematical thought. Still others relate mathematics to the material world. The biographical sketches will make it clear that great mathematicians, though extraordinary in their abilities, are people like other human beings and not practitioners of black magic.

Hollis R. Cooley

INTRODUCTION

AS THE CONTENTS of this volume will make clear, the
universe, from the most distant nebulae to the smallest units
of subatomic matter, seems to be ruled by mathematical law.
It has been aptly stated that God is a mathematician, and
certainly mathematics is a subject which has universal applications. It is used by the corner grocer to tot up bills and
by the atomic physicist to formulate hypotheses. These two
examples, though they may seem as far apart as the poles,
nevertheless have one thing in common—they are both concerned with the *uses* of mathematics. But there are other
aspects of the subject. There are mathematicians who speak
with condescension about such applications and who feel
that the highest type of mathematics—so-called "pure" mathematics—has no practical value whatever.

Members of this group—among whom have been numbered some of the greatest mathematicians—find in the subject "a beauty cold and austere." They take pride and pleasure in the fact that the discipline they follow is a world unto
itself, an exercise in pure reason whose mysteries, though
incidentally they may prove of value for the mundane
world, are important because they represent *truth* in its most
rarified form.

In this volume, a few of the best expressions of both the
practical and the abstract points of view are presented. Within the limitations of space available, the reader is given some
indication of how mathematics has developed and how some
of the greatest figures, from Archimedes to Einstein, have
shaped its course. He glimpses the multitudinous branches
which have grown from roots planted in prehistory. He is
offered an insight into the diversity, the profundity and the
cultural value of this branch of knowledge. George Sarton

xi

has called the history of mathematics "the kernel of the history of culture . . . the skeleton which supported and kept together all the rest" of the sciences. Yet mathematics is too often neglected by even the most intelligent men and women once they have completed their formal educations.

Only a multivolume work could do full justice to the subject. The literature of mathematics is of extraordinary richness and interest—a fact which may surprise those who consider the very language which it employs as difficult of comprehension as Choctaw. But from this mass of material there does emerge an understandable pattern and it is the purpose of this volume to present it in a comprehensible and even exciting manner. Here, in the words of some of the outstanding writers on the subject, is a picture of what mathematics is, what mathematicians are like, what inspiration—akin almost to religion—it offers.

The book is addressed to laymen, and the knowledge of mathematics which is required for an understanding of its contents is minimal. Scholarly apparatus, such as footnotes and references, has been omitted. In some cases, articles have been presented in somewhat abbreviated form.

MATHEMATICS is one of the books in the series entitled THE NEW YORK UNIVERSITY LIBRARY OF SCIENCE. Other volumes, dealing with individual sciences, are in preparation. The series as a whole will encompass much of the universe of modern man, for that universe has been shaped in greatest measure by science, the branch of human activity whose name is derived from the Latin *scire*—to know.

MATHEMATICS

I. Biography of MATHEMATICS

I. Biography of Mathematics

In his article on Big Numbers which appears later in this
book, George Gamow describes the numbers systems of
primitive tribes who can count only to three; and of more
sophisticated peoples who evolved systems of pairing objects
in different collections to compare the quantities in each.
Such beginnings are observable today among the Hottentots,
but in general are hidden in the mists of prehistory. A primitive mathematics—probably beginning with the use of the
fingers in counting—certainly existed before the time of our
earliest written records. In the fourth millenium B.C., to
quote George Sarton, author of the article which follows, "the
Sumerians had already some kind of 'position' concept in the
writing of numbers, and had learned to treat submultiples in
the same way as multiples," which was later lost for millenia.
Babylonian tablets of as early as 2000 B.C. show a surprising
knowledge of mathematics—a numbers system using 60 as a
base, fractions, astronomical measurements, and algebraic exercises. As early as 3500 B.C., the Egyptians were acquainted
with numbers in the millions. Even before the Rhind Papyrus,
a manuscript dating from around 1700 B.C., they had used
arithmetic and geometry in constructing and orienting the
pyramids. There are trace influences of Babylonian methods in
Egyptian mathematics and the Babylonians can probably lay
claim to precedence. The Babylonians also seem to have made
some discoveries, such as the Pythagorean theorem, which
were later ascribed to the Greeks. Yet it is the Greeks who
must be credited with the founding of mathematics as a scientific discipline.

The achievement which entitles them to this distinction is
the invention of a system of certain "self-evident" (later not so
self-evident) axioms, from which, through a series of logical
deductions, certain mathematical proofs could be developed.
Obviously, even at this early date, mathematics and logic

1

were inextricably entwined, and throughout history this marriage has been undissolved.

With the tool or method which the Greeks had invented, and with the empirical knowledge they had inherited, Greek mathematics flared swiftly into a complex structure. A school whose major interest was geometry was founded around the seventh century B.C. and lasted for nearly a millenium. Among its earliest practitioners were Thales of Miletus, and Pythagoras. The former is said to have predicted an eclipse in 585 B.C., and to have taken the first steps in deductive geometry. The latter is said to have discovered the law of musical intervals, to have sacrificed an ox in celebration of his proof of the theorem named for him (but see above), and to have invented the pseudoscience of numerology, in which virtue, sex, or metaphysical attributes were linked with certain numbers or kinds of numbers—a doctrine which succeeding generations of fortune-tellers have made use of. Pythagoras was a shadowy figure for whose achievements, though credited to him, whole schools were actually responsible. These schools, which bore his name, were conducted as secret societies. They investigated the properties of areas in a plane, and of most of the regular solids. Their studies had the distinction of being "philosophical"— done purely for the acquisition of knowledge.

There were other schools. The Eleatics of Italy, for example, of whom Zeno was one, propounded paradoxes which were to baffle mathematicians until fairly recent times. But the main thread of mathematical development took place in Athens during the Golden Age, and at its related school of Cyzicus. Among the leaders of these schools were Hippocrates of Chios (not to be confused with the physician of Cos), who used the method of reductio ad absurdum; Eudoxus, founder of the school of Cyzicus and teacher of Menaechmus, about whom Eric Temple Bell writes in From Cyzicus to Neptune (p. 253); and Plato and Aristotle, both of whom were interested in the connection between mathematics and rigorous reasoning.

Menaechmus was the teacher of Alexander, whose conquest of Greece in 330 B.C. was the beginning of a shift to Alexandria of the line of development. But the spirit and practice of mathematics remained Greek in essence. The Alexandrian mathematician whose name is most familiar to us is of course Euclid. His personality is nebulous, but he wrote a textbook

which has survived for a longer period than the New Testament and has probably been read by a comparable number of
people. His work was one of codification—it drew on discoveries made over a period of hundreds of years—but its
clarity of presentation was superb. A later, more original member of the Alexandrian school was Apollonius, "the great
geometer" who specialized in the study of conic sections. An
equally noteworthy figure was Hipparchos the trigonometer
and astronomer. But Archimedes was certainly the leading
mathematician of antiquity. Read against the background of
mathematics which preceded him, his biography takes on both
interest and importance.

This biography was written by George Sarton, who was
probably the foremost historian of science of modern times.
Born in Belgium in 1884, he came to the United States in
1916 and became Professor of the History of Science at
Harvard. He died in 1956. Sarton was the author of a number
of works including the monumental Introduction to the History of Science. A man of broad humanitarian outlook, he was
perhaps the foremost exponent of the interrelationship of all
the sciences, and in turn, of their interrelationship with every
other aspect of human culture. Even in this abbreviated life
of Archimedes, we can observe clearly his encyclopedic learning, his meticulous research, and his understanding of the
processes of scientific creation.

ARCHIMEDES

GEORGE SARTON

WHEREVER GREEK COLONIES were established, there were
definite possibilities of scientific advance, the outstanding
[example] in the third century being that of Archimedes of
Syracuse. From the sixth century on, the two leading cities
in [the Western Mediterranean] were Carthage in the
Semitic Empire and Syracuse in the Greek one.

Carthage was the older settlement. It had been founded by a Tyrian swarm as early as 814, and we all know the first Queen, Dido, immortalized in the *Aeneid*. It soon became the main colony of its kind, to such an extent that one ceased to speak of Phoenicians and spoke instead of Carthaginians.

Syracuse was founded eighty years later than Carthage, in 734, on the southeast coast of Sicily. Thanks to its magnificent location and to the genius of its Corinthian founders, it soon became the most important city not only of Sicily but of the whole of Magna Graecia. It was bound to antagonize Carthage and the perils of war caused the establishment of a dictatorship from 485 on. The struggle between Syracuse and Carthage continued until the Romans, taking advantage of a pro-Roman party, besieged the city and took it in 212.

When the Roman general Marcellus besieged Syracuse, the difficulty of his task was greatly increased by the resourcefulness of an engineer named Archimedes, who was killed during the sack of the city in 212. According to the legend, Archimedes had invented various machines for defensive purposes, catapults, ingenious hooks, and also concave mirrors by means of which he deflected the sun rays and set the Roman ships on fire. The story is told that a Roman soldier came upon him while he was absorbed in the contemplation of geometric figures drawn up on the earth. Archimedes shouted, "Keep off," and the soldier killed him. The account of the inventions by which he tried to save his native city fired the imagination of people not only during ancient and medieval times but even as late as the eighteenth century, and he was generally thought of as a mechanical wizard. For example, Gianello della Torre, clockmaker to Charles Quint, was called "the second Archimedes" and as late as the eighteenth century, the inventor, Christopher Polhem, was called "the Swedish Archimedes." That is as silly as if we were to call Edison "the American Archimedes." The absurdity of such nicknames is obvious as soon as one realizes that Archimedes, though he may have invented various machines and gadgets was primarily a mathematician, the greatest of antiquity and one of the very greatest of all times.

Plutarch had already remarked that Archimedes himself did not think much of his practical inventions. Yet, it is certain that Archimedes' fame was based for many centuries not upon the immortal achievements explained in his own works but upon the legends that clustered around his name. These legends had a core of truth; he did invent machines, such as compound pulleys, an endless screw, a hydraulic screw, an orrery, burning mirrors, but these activities were secondary and marginal.

The only fact of his life that can be dated with certainty is his death during the sack of Syracuse in 212. As he was said to be 75 years old in that year, that places his birth c. 287. He was the son of the astronomer Pheidias; his early interest in astronomy and mathematics was thus natural enough. He was a kinsman and friend of Hieron II, king of Syracuse, and of the latter's son and successor, Gelon II. According to Diodoros of Sicily, he spent some time in Egypt and that is very plausible. Alexandria was then the center of the scientific world; Archimedes had no equal in Syracuse, and he would naturally wish to visit the museum and exchange views with the great mathematicians who were flourishing in its neighborhood.

One more story. Archimedes requested his friends to engrave a mathematical diagram on his tombstone. The diagram (or was it a tridimensional model?) represented a cylinder circumscribing a sphere. We know this through Cicero, who, when he was quaestor of Sicily in 75 B.C., discovered Archimedes' tomb in a ruined state, restored it, and described it. The tomb has disappeared and its exact location is unknown.

Now that we know the man Archimedes as much as is possible, let us consider the extant works that have immortalized him.

Archimedes lacked the encyclopedic tendencies of Euclid, who tried to cover the whole field of geometry; he was, on the contrary, a writer of monographs of limited scope, but his treatment of any subject was masterly in its order and clearness. As Plutarch remarked in his life of Marcellus, "It is not possible to find in geometry more difficult and troublesome questions or proofs set out in simpler and clearer propositions."

The longest of all Archimedes' writings is a treatise on the *Sphere and cylinder* in two books, the Greek text of which covers (in Heiberg's edition) not more than 114 pages. In that treatise he proves a number of propositions, such as the one to which he himself attached so much value that he ordered the diagram relative to it to be engraved on his tombstone, and also the one which every schoolboy knows, that the area of the surface of a sphere is four times that of one of its circles ($4 \pi r^2$). The treatise begins in Euclidean fashion with definitions and assumptions. For the determination of surfaces and volumes, he uses the method of exhaustion, very skillfully and rigorously. He solved the problem, "To divide a sphere by a plane into segments the volumes of which are in a given ratio," and similar ones.

His second treatise in order of length (100 pages in Greek) is the one on *Conoids and spheroids,* dealing with paraboloids and hyperboloids of revolution and the solids formed by the revolution of ellipses about their major or minor axis. The third (60 pages) is devoted to *Spirals.* This third treatise summarizes the main results of the two preceding ones and hence is also the third in chronologic order. The spiral he dealt with is the one that is called to this day the "Archimedean spiral" and which he defined as follows: "If a straight line of which one extremity remains fixed be made to revolve at a uniform rate in a plane until it returns to the position from which it started, and if, at the same time as the straight line revolves, a point moves at a uniform rate along the straight line, starting from the fixed extremity, the point will describe a spiral in the plane." That clear definition would still be used today and would lead to the equation $r = a\theta$, wherein a is a constant. (There are, of course, no equations in Archimedes, nor in any other ancient text; our equations hardly date back to the second half of the sixteenth century.) He finds various areas

bounded by it, and what we would call the constancy of its
subnormal $(=a)$. His ability to obtain those results without
our analytical facilities is almost uncanny.

His fourth treatise, on the *Quadrature of the parabola*,
is much shorter (27 pages), but deals with a single problem.

These four geometric treatises constitute the main bulk
of Archimedes' available works. His other geometric treatises are much shorter and less important. The *Measurement of the circle* (perhaps a fragment of a larger treatise)
leads to a good approximation of π, namely, $3\frac{1}{7} > \pi > 3\frac{10}{71}$
$(3.142 > \pi > 3.141)$. Archimedes had obtained that result
by comparing the areas of two regular polygons of 96
sides, inscribed in and circumscribed about the same circle.

This enumeration is more than sufficient to reveal the
incredible depth and ingenuity of Archimedes' geometric
thought. Not only did he ask questions that were original
and obtain results that were almost unthinkable in his time,
but he used methods that were rigorous and unique. For
example, he accomplished quadratures of curvilinear plane
figures and the quadrature and cubature of curved surfaces. By means of a method equivalent to integration, he
measured the areas of parabolic segments and of spirals,
the volumes of spheres, segments of spheres, and of segments of other solids of the second degree. It is foolish
to speak of him as a forerunner of the inventors of analytic
geometry and of the integral calculus, but the very fact
that such claims could have been made for him is highly
significant. When one bears in mind that he had formulated
and solved a good many abstruse problems without having
any of the analytic instruments that we have, his genius fills
us with awe.

Archimedes' work in arithmetic and algebra was less bulky
and less original. Was he at all acquainted with Babylonian
methods, I wonder? He might have heard of them during
his stay in Alexandria; it would not have been necessary for
him to hear much, the most meager suggestion would suffice
to excite his mind. At any rate, it is not possible to recognize
definitely Babylonian elements in his works.

Archimedes had been impressed by the inherent weakness of the Greek numerical system, whether it be expressed with words or with symbols. That weakness is one of the paradoxes of Greek culture; the leading mathematicians of antiquity had to be satisfied with the worst numerical system, the very basis of which was hidden by inadequate symbols. His own genius was wanting in this case, for instead of inventing a better system (that was the true solution), he tried to justify Greek numerals by showing that they were sufficient to designate the very largest numbers. Of course, every number system, however poor, could be justified in the same manner. He explained his views *ad hoc* in a treatise entitled the *Psammites* (*Arenarius, Sand reckoner*), dedicated to King Gelon, wherein an extremely large number is introduced in a very original way. "How many grains of sand could the whole universe hold?" It is clear that that question is double, for one must first determine the size of the universe; this being done, if one knows how many grains are contained in a unit of space it is easy to calculate how many the whole universe will be able to hold. It is easy provided we have the necessary number words. In the decimal system, the question would not arise, because if one understands the meaning of 10^0, 10^1, 10^2, there is no difficulty in understanding 10^n, irrespective of the size of n. Archimedes' solution was more complicated. The numbers from 1 to 100 million (10^8) formed his first order, those from 10^8 to 10^{16} the second order, and so on; those of the 100-millionth order ending with the number $10^{8 \cdot 10^8}$. All of these numbers form the first period; a second period may be defined in the same way, also a third, and so on up to the 10^8th period, ending with the number $(10^{8 \cdot 10^8})^{10^8}$. The decimal expression of the last number of the 10^8th period would be 1 followed by 80,000 million million zeros. The number of grains of sand in the universe is relatively small, less than 10^{63}.

This aspect of Archimedes' genius is curious; instead of thinking out a numerical system that would be of use in practical life, he indulged in the conception of immense numbers—a conception that is philosophical rather than purely mathematical.

We now come to something that is perhaps even more remarkable than Archimedes' geometric investigations, namely, his creation of two branches of theoretical mechanics, statics and hydrostatics. Two of his mechanical treatises have come down to us, the *De planorum aequilibriis* and the *De corporibus fluitantibus,* both composed in Euclidean style, divided into two books, and of about equal length (50 pp. and 48 pp.). They both begin with definitions or postulates, on the basis of which a number of propositions are geometrically proved.

The first, on the *Equilibrium of planes,* begins thus:

I postulate the following:
1] Equal weights at equal distances are in equilibrium, and equal weights at unequal distances are not in equilibrium but incline toward the weight which is at the greater distance.
2] If, when weights at certain distances are in equilibrium, something be added to one of the weights, they are not in equilibrium but incline toward that weight to which the addition was made.

After a few steps, he is then able to prove that "two magnitudes, whether commensurable or not, balance at distances reciprocally proportional to them." The distances to be considered are the respective distances of their centers of gravity from the fulcrum. Therefore, the end of Book I explains how to find the centers of gravity of various figures —parallelogram, triangle, and parallel trapezium—and the whole of Book II is devoted to finding the centers of gravity of parabolic segments.

The treatise on *Floating bodies* is based on two postulates, the first of which is given at the beginning of Book I and the second after Prop. 7 (out of 9). They read:

Postulate 1.

"Let it be supposed that a fluid is of such a character that, its parts lying evenly and being continuous, that part which is thrust the less is driven along by that which is thrust the more; and that each of its parts is thrust by the fluid which is above it in a perpendicular direction if the fluid be sunk in anything and compressed by anything else."

Postulate 2.

"Let it be granted that bodies which are forced upwards in a fluid are forced upwards along the perpendicular [to the surface] which passes through their center of gravity."

On the basis of Postulate 1, he proves that "the surface of any fluid at rest is a sphere the center of which is the same as that of the Earth." The main propositions of Book I, Props. 5–7, are equivalent to the famous Archimedean principle according to which a body wholly or partly immersed in a fluid loses an amount of weight equal to that of the fluid displaced. It has often been told that he discovered it when he was aware of the lightness of his own body in the water, and that he ran out of the bath shouting with joy "Heureca, heureca" (I have found it). This enabled him to determine the specific gravity of bodies and to solve the "problem of the crown." A golden crown made for King Hieron was believed to contain silver as well as gold. How great was the adulteration? The problem was solved by weighing in water the crown itself, as well as equal weights of gold and silver.

It would seem that Archimedes wrote at least one other mechanical treatise, wherein he solved the problem "how to move a given weight by a given force," and proved that "greater circles overcome lesser ones when they revolve about the same center." This recalls his legendary boast to King Hieron: "Give me a point of support [a fulcrum] and I shall move the world." In order to convince the king, he managed to move a fully laden ship with no great effort by the use of a compound pulley.

This brings us back to the mechanical inventions for war and peace which impressed posterity so profoundly that

Archimedes' theoretical achievements were overlooked. The magnitude of his work in pure statics and hydrostatics can be appreciated in another way. Remember that Aristotelian and Stratonian physics was absolutely different from physics as we understand it today. The first physical sciences to be investigated on a mathematical basis were the rudiments of geometric optics (by Euclid and others) and, more deeply, two branches of mechanics, statics and hydrostatics. This was done by Archimedes, who must be called the first rational mechanician. Nor was there any other at all comparable to him until the time of Simon Stevin and Galileo, who were born more than eighteen centuries later!

Few mathematicians explain their method of discovery and, therefore, their accounts are often tantalizing, for one cannot help wondering, "How did he think of that?" Their reticence may be due to a kind of coquetry but, in most cases, it is simply the fruit of necessity. The first intuition may be vague and difficult to express in scientific terms. If the mathematician follows it up, he may be able to find a scientific theory, but his way to it is tortuous and long. To describe the discovery in historical order would be equally long and tedious. It is simpler to explain it logically, dogmatically, after having thrown out everything irrelevant. The new theory then looks like a new building, after the scaffoldings and all the auxiliary construction have been taken away, without which the building could not have been erected.

One cannot read Archimedes' complicated accounts of his quadratures and cubatures without saying to oneself, "How on earth did he imagine those expedients and reach those conclusions?" Eratosthenes must have asked the same question, not only of himself but of Archimedes. The point is that the conclusions were reached intuitively and roughly before their validity was proved, or before it was possible to begin such a demonstration.

We must still say a few words about Archimedes' work in the fields of astronomy and optics. He wrote a (lost) book on *Sphere-making* describing the construction of an orrery to show the movement of Sun, Moon, and planets; the orrery was precise enough to foretell coming eclipses of the Sun and Moon. In the *Sand reckoner,* he described the simple apparatus (a diopter) that he used to measure the apparent diameter *d* of the Sun. According to Macrobius, Archimedes determined the distances of the planets.

His interest in optics is proved by another lost book, *Catoptrica,* out of which Theon of Alexandria quoted a single proposition: objects thrown into water look larger and larger as they sink deeper and deeper.

Considering the history of Greek astronomy and optics, it is not surprising that Archimedes paid some attention to those subjects. During his stay in Alexandria, he had discussed them with the disciples of Euclid and Aristarchos. Nevertheless, his own main interest was mathematical and it is admirably illustrated in the books that have come down to us.

Archimedes was a direct descendant of the Greeks of the Age of Pericles, who were, as we have said, in the mainstream of mathematical history. There were, however, parallel currents which were not to join until long after Archimedes was slain. The Roman conquerors had a debilitating influence on the science. And the Arabs, in one of civilization's greatest tragedies, destroyed the great library and museum at Alexandria. Yet the Arabs also made a positive contribution. They had a considerable system of mathematics, based in part on that of the East, and it was they who united the strands of Hindu and Chinese mathematics with those which stemmed from Athens. The Greeks, with all their inventiveness and intuition in geometry, had continued to struggle with an extremely cumbersome arithmetic. This system was all but useless for anything but the most primitive calculation. The Hindus had a far more flexible and functional numbers system and with it one of the foundation stones of mathematics, the zero. We are almost completely ignorant of the early history of Hindu mathematics. They had knowledge of the value of $\sqrt{2}$ and of π, and their mathematics was closely linked to such diverse matters as puzzles for entertainment and astronomy. In the seventh century A.D., the name of Brahmagupta, and in the twelfth, of Bhaskara, are recorded. They were arithmeticians, and Bhaskara is supposed to have made a careful formulation of the decimal system.

Arabic notation was so obviously superior that it gradually superseded the Roman, but it was centuries before the ease of calculation which it provided came into common use. Even in the sixteenth century, Melancthon, Luther's associate, who was also a professor in search of students, stated that "mathematics is not as difficult as ordinarily supposed and even division may be mastered by diligence."

The Arabic influence began during the Moorish invasion of Europe in the seventh and eighth centuries. Around 1100 A.D., it was fostered by the crusades and by study at the universities, which were growing in importance. But the pall of the Middle Ages still hung over all intellectual activity, and not until the latter half of the fifteenth century did mathematics begin to share in the glorious activity of all the arts and sciences during the Renaissance. Until that date, this Renaissance was being slowly prepared. Once in flower, it was to continue, unlike that of the Greeks, without interruption. The development of mathematics and the sciences were of course inextricably bound together during the period. Mere mention of the names of Copernicus, Kepler, and Galileo is enough to establish the relationship among astronomy, physics, and mathematics. This was also the great era of discovery and the needs of explorers and map makers led to improvements in trigonometry and more accurate star tables. Napier invented logarithms, Descartes analytical geometry, Fermat the calculus of probabilities. The latter's greatest contribution was his development of the basis for modern numbers theory. Apology is perhaps in order for dismissing with a phrase men whose lives were monuments in the history of science.

In one respect, mathematics was favored over physics and astronomy—it was fostered by the Church, which saw in it no obstacle to faith. It was to an extent hindered by the secrecy with which mathematicians hedged their discoveries and which resulted in violent jealousies and priority feuds. Nevertheless,

mathematics was flourishing to such a degree that Isaac Newton, the greatest of the craft, was to say, commenting on his
own work in mathematics, as well as in physics and astronomy,
"If I have seen farther than others, it is by standing on
the shoulders of giants." The following biography is from
Hooper's Makers of Mathematics.

NEWTON

ALFRED HOOPER

I do not know what I may appear to the
world; but to myself I seem to have been
only like a boy playing on the seashore, and
diverting myself in now and then finding
a smoother pebble or a prettier shell than
ordinary, whilst the great ocean of truth lay
all undiscovered before me.
    —BREWSTER'S MEMOIRS OF NEWTON.

IN OCTOBER, 1661, a youth between eighteen and nineteen years of age traveled to Cambridge and found his way
to Trinity College. The previous June, the Master of Trinity
had admitted him as a member of the great college founded
more than a century previously by Henry VIII. On July 8th
he had been admitted as a student-member of the University of Cambridge. Now he had "come up" for the October
term and was about to commence his life there as an unknown undergraduate.

As he entered the courtyard with its statue that he was to
find was in memory of Thomas Neville, the Master of Trinity
from 1593 to 1613, and saw for the first time the college
chapel built in Queen Mary's reign, he little could have
dreamed that one day his own statue would be placed in
that chapel, and that one of the world's greatest poets would
describe that statue as

The marble index of a mind forever
Voyaging through strange seas of thought alone.

He would undoubtedly have been annoyed and embarrassed had he known that Pope was to rhapsodize extravagantly

Nature and Nature's laws lay hid in night:
God said "Let Newton be!" and all was light.

Never having read a book on mathematics, he most definitely could not have conceived the possibility that within eight years of that October day he would be appointed a professor of mathematics in the University of Cambridge, and that he would write a book on science and applied mathematics which, more than a century later, was to be described by the profound French mathematician Laplace (usually very sparing in his tributes to others) as assured of "a preeminence above all other productions of human genius."

Isaac Newton was born on Christmas day 1642 (the year of Galileo's death, and the year when civil war broke out between Charles I and the English Parliament) in the little village of Woolsthorpe, Lincolnshire, some six miles from the town of Grantham. He never knew his father, for the latter, a small farmer with a reputation for extravagance, had died before his son was born. The boy was so frail at birth that he was not expected to live. Thus he was given his first of many opportunities of astonishing others. As in the case of all these other opportunities, he seized on this first one to the full, living to his eighty-fifth year, when he still had perfect vision and a thick head of silvery hair — his hair turned prematurely gray when he was thirty.

His childhood's frailty prevented him from joining in the rough and tumble of village schoolchildren. Instead, he amused himself by making mechanical toys: little working models of such things as waterwheels and windmills, a tiny carriage moved by its rider, and ingeniously constructed kites. The mechanical skill and ingenuity he thus developed were to be very useful to him later in life, when making practical experiments in connection with lenses and the study of the properties of light. The joy of creating things thus tasted early in life never deserted him; he passed from the creation of mechanical objects to that of intellectual concepts. Thus it came about that he never felt the need to have others "amuse" and "entertain" him.

His interests were overwhelmingly intellectual; he never married; he cared nothing for dress. Like Archimedes, he possessed the power of concentration to a remarkable degree. His mind would close on a problem like a steel trap, and nothing was allowed to divert it from its goal. Meals would be ignored, and, on getting up in the morning he would forget to dress and would be found hours later sitting in his bedroom oblivious to everything save the problem in which he was absorbed. Although Nature had been so lavish in her mental equipment, he once came near to ruining her gifts by his neglect of food and his reluctance to spend precious hours in sleep. In the Michaelmas Term 1684 he had given a course of lectures at Cambridge on the laws of motion. By 1685 he had incorporated these lectures into what was to be the first of three books which were to constitute his world-famous masterpiece, the *Principia*. This book gave the world his law of universal gravitation, which, as Dr. F. J. Cajori said in his *History of Mathematics* "envelops the name of Newton in a halo of perpetual glory."

The second book was completed in the summer of 1685, and the third in the incredibly short time (in view of its contents) of nine months. The years of intense concentration and labor involved in the production of these great works undermined his strength and some years later brought on what would now be described as a near nervous breakdown. Fortunately he made a complete recovery and the very next year after his illness he showed that his powers were unimpaired by solving the problem of the *brachisto-chrone* (*brachistos*, "shortest," *chronos*, "time") or curve of quickest descent. This problem had been issued as a challenge to other mathematicians by Johann Bernoulli, who had coined the Greek word *brachistochrone*. Newton received the problem on January 29, 1697, and gave the solution (a cycloid) on January 30, 1697. Again, in 1716, Leibnitz issued an extremely difficult problem. Newton, aged seventy-four, solved it in five hours. Nevertheless, after completing the *Principia*, a distinct change took place in Newton. He produced very little original mathematical or scientific work. He allowed the cordial relationship that had existed between himself and Leibnitz, for example, to degenerate into a shamefully distressing state of animosity during all the years when, as President of the Royal Society, a word from him would have saved the situation. Possibly his illness had deeper effects than those that appeared on the surface.

It is not surprising to learn that during the years when he was exercising intense concentration on mathematical and scientific matters he was at times forgetful of everyday things. On one occasion, it is said, while leading his horse up a steep hill, he turned his mind to some problem. Some considerable time later he was puzzled to find the bridle still in his hand, but no horse attached to it. On another occasion—one of the few occasions when he entertained friends—he went from the room to fetch more wine for his guests. After a long and trying interval, his thirsty guests were driven to investigate their host's prolonged absence. They found him absorbed in some problem or other. Without doubt, the thought of wine would bring to Newton's mind Kepler's use of infinitesimals rather than the qualities usually associated with the "vinous beverage."

At the very beginning of his career he discovered that mathematicians and scientists were quick to question and criticize any departure from habitual practices. At first Newton took great pains to answer their criticisms with patience and courtesy, and to try to clear up their misunderstandings of his work. But as their criticisms continued, he became hypersensitive on the subject, and bitterly resented the hours—wasted hours, in his opinion—that were consumed in replying to his critics. "Philosophy [science] is such an impertinently litigious lady," he remarked in a letter to his good friend, the eminent astronomer Edmund Halley, "that a man has as good be engaged in lawsuits, as have to do with her." Halley's reply, dated June 29th, shows the tact and patience with which he handled the sensitive genius. He said, "I am heartily sorry that in this matter, wherein all mankind ought to acknowledge their obligations to you, you should meet with anything that should give you unquiet. . . . I am sure that the Society [Royal Society] have a very great satisfaction, in the honour you do them, by the dedication of so worthy a treatise [the first two books of the *Principia*]. Sir, I must again beg you, not to let your resentments run so high as to deprive us of your third book." The world has to thank Halley, not only for bearing the cost of printing the *Principia* and for correcting all the proofs, but also for so gently and tactfully urging Newton to complete his great work.

Newton's life falls into three distinct sections: the first covers his boyhood in Lincolnshire; the second, his life at Cambridge from 1661 to 1696; the third, his work as a highly paid government official from 1696 to his death in 1729.

After attending two small schools in hamlets close to Woolsthorpe, he was sent, at the age of twelve, to the grammar school at Grantham. Newton later admitted that his early days at Grantham school were far from industrious. He stood low in his class, and, doubtless owing to his physical frailty, he seems to have suffered from an inferiority complex until the day when he was goaded into a fight with a bully. Newton's biographer, Sir David Brewster, says that Newton got the better of the fight, and, dragging the bully by the ears, pushed his face against the wall. This story seems a little too good to be true, but there is no question that from this time onward Newton's work showed great and rapid improvement. Brewster attributes this sudden development of mental concentration to Newton's determination to vanquish the bully in class. Not only was Newton eventually successful in doing this, but he rose to be head of the school.

When he was fifteen, however, his mother, who had remarried in 1645 and had become a widow for the second time in 1656, withdrew him from the school, feeling that it was time he began to learn how to cultivate his father's small farm. Farming, however, did not appeal to the boy.

When on the farm, he preferred to sit under a tree reading a book or carving a model to preventing the sheep and cattle from straying, while in later years he told a friend that on Friday, September 3, 1658, when a great tempest swept over England as Oliver Cromwell passed away, he occupied himself in trying to measure the force of the gale by leaping, first with the wind and then against it, and then comparing these measurements with the distance he could jump on a calm day. It was activities such as these (when a farmer's mind ought to be occupied in protecting his stock and equipment) that caused his mother to realize that he was unfitted for farming. Fortunately for mankind, she had the good sense to send him back to Grantham Grammar School, where he was prepared for entrance to Cambridge University on the advice of his uncle, the Reverend W. Ayscough, the rector of a neighboring parish.

Very few details have come down to us of his undergraduate days at Cambridge. The best we can do is to try and piece together some notes made by him many years later.

In his first term, that is, between October and Christmas 1661, he went to the little village of Stourbridge, near Cambridge, in order to visit the fair that was held there every year, at one time the greatest fair in England. Here he bought a book on the stars, but found he was unable to understand it on account of his ignorance of geometry. So he bought an English edition of Euclid's *Elements*. He found the contents so "self-evident" (to quote his own words) that he put aside the *Elements* as a "trifling book"! He therefore got hold of a copy of Descartes' *Geometry* and, after a hard struggle, managed to master it. In 1664 he sat for a scholarship at Trinity, to which he was elected on April 28th of that year, despite the fact that one of his examiners, Dr. Barrow, the first occupant of the Lucasian Chair of Mathematics, reported adversely on his knowledge of Euclid's *Elements*, a verdict which is not surprising, in view of the extent of his acquaintance with the "trifling book."

This led Newton to study the *Elements* with care, and thus come to realize that the book was no trifling matter, as he had supposed. He was later to make a masterly use of Euclid's geometry in order to give to the world his mathematical explanation of universal gravitation. He had himself reached his conclusions by means of his own invention, the calculus; since, however, he knew that other mathematicians were ignorant of his invention—and would doubtless raise objections to it and thus deny the truth of conclusions based on it—he recast his arguments in geometric form, as we shall see. Thus, Barrow's criticism of the young candidate for a Trinity scholarship was to bear good fruit in the end.

In January 1665, he took the degree of Bachelor of Arts but was later in that year forced to leave Cambridge owing to the plague. While the college was closed, Newton went to Woolsthorpe, and there it was that he began to think about the fundamental principles of his theory of gravitation.

It was here, in 1666, that the well-known legend of Newton and the apple arose. Tradition holds that the idea of gravitation was suggested to Newton by the fall of an apple. Certainly, the supposed tree from which it fell was kept standing until a gale destroyed it in 1820. But the earth's gravitation was an accepted scientific fact long before Newton's time; his genius lay in developing the law of universal gravitation.

Living in the peaceful countryside where he had once feared he would have to spend his life as a farmer, he invented his "fluxional calculus," which was to enable him to build up his law of universal gravitation, and was to give mathematicians and scientists their most powerful weapon.

From "the beginning of the year 1664, February 19th" Newton was observing the heavens. He found he must have some mathematical tool at his disposal if he was to find an explanation for Kepler's laws. So he developed the "calculus" which did for the "mathematics of motion and growth" what Descartes' Geometry had done for the geometry of the Greeks. Like Descartes, he built on the foundations laid by others, and produced a generalized, far-reaching branch of mathematics, whose tremendous consequences he could little have foreseen.

> (*Appreciation of the contribution of Newton to the development of mathematics is of course impossible without an understanding of the calculus, which Newton and Leibnitz invented independently and which the former used to work out his "system of the world." For an exposition of the method and meaning of the calculus, the reader is referred to "Change and Changeability—the Calculus" by Kasner and Newman which appears on page 183.—Eds.*)

On October 1, 1667, he was made a Fellow of Trinity College and the following year he took the degree of Master of Arts.

In the summer of 1669 he handed Barrow a paper he had drawn up and in which he had partly explained his principle of fluxions. Barrow's admiration of this paper was so great that he described Newton to another Cambridge mathematician, John Collins, as of "unparalleled genius." Unfortunately, Newton was too modest to follow Barrow's advice and publish this paper. Had he done so, it would have prevented years of undignified squabbling between Newton's followers and those of Leibnitz.

Shortly after receiving this evidence of Newton's mathematical genius, Barrow, who wished to devote his time to the study of theology, resigned his professorship of mathematics, and on his strong recommendation, Newton, then only twenty-seven years of age, was appointed to succeed him.

In those days the Lucasian Professor of Mathematics was required to lecture only once a week during term time, on some mathematical subject or on astronomy, geography, optics (the laws of light), or statics, and to give up two hours a week for consultation with students. Newton chose optics for his first lectures, in which subject he had already made far-reaching researches and discoveries. The results of his researches, however, were for some years known only to his Cambridge audiences. When, however, he was elected a Fellow of the Royal Society in 1672, he sent that body a long paper based on his lecture notes, and, in this way, his work in connection with the composition of light became known to scientists throughout Europe.

Between the years 1673 and 1683 his lectures at Cambridge were on the subject of algebra, particularly with regard to the theory of equations. His lecture notes were put into book form and printed in 1707; they deal with many important advances in this subject, especially in connection with the so-called "imaginary" roots of certain equations.

In 1684 Newton was paid a visit by his friend Edmund Halley which was to have momentous consequences. Halley, Hooke, Huygens and Wren had been engaged in trying to find an explanation for the laws empirically discovered by Kepler regarding the movements of the heavenly bodies. Halley explained that their investigations were held up by their inability to apply Kepler's laws in order to determine the orbit of a planet. Newton immediately told Halley that, some five years before, he had proved that the orbit was an ellipse. He was not able to put his hand on the paper in which he had made the calculation in 1679, so he promised to work it again for his friend. This promise evidently led him to return once more to the subject of universal gravitation during the summer vacation of 1684, for his lectures during the Michaelmas term of that year dealt with this subject. Halley visited him again in the middle of this term, and studied his manuscript lecture notes. These notes, entitled *De Motu Corporum* ("Concerning the Movement of Bodies") are to be seen today in Cambridge University Library. Halley urged Newton to publish them, but had to be content with a promise that they would be sent to the Royal Society, which promise Newton kept early the following year. Thanks to the tactful pressure exerted by Halley, which we mentioned earlier in this chapter, Newton now became deeply engrossed in the whole problem of gravitation. In 1685 he was able to prove that the total attraction of a solid sphere on any mass outside that sphere could be considered as if concentrated in a single point at its center. No sooner had Newton proved this superb theorem—and we know from his own words that he had no expectation of so beautiful a result till it emerged from his mathematical investigation—than all the mechanism of the universe at once lay spread before him. . . . In his lectures of 1684, he was unaware that the sun and earth exerted their attractions as if they were but points. How different must these propositions have seemed to Newton's eyes when he realized that these results, which he had believed to be only approximately true when applied to the solar system, were really exact! Hitherto they had been true only in so far as he could regard the sun as a point compared to the distance of the planets, or the earth as a point compared to the distance of the moon—a distance amounting to only about sixty times the earth's radius—but now they were mathematically true, excepting only for the slight deviation from a perfectly spherical form of the sun, earth, and planets. We can imagine the effect of this sudden transition from approximation to exactitude in stimulating Newton's mind to still greater efforts. It was now in his power to apply mathematical analysis with absolute precision to the actual problems of astronomy. [Dr. Glaisher's address on the bicentenary of the publication of the *Principia;* quoted in W. W. R. Ball's *History of Mathematics* (Macmillan).]

The efforts to which Newton's mind was now stimulated were so immense that by April, 1686, he sent the first book of the *Principia* to the Royal Society; the second book by the summer of the same year; the third book he had completed in manuscript form by 1687. The whole work was published at Halley's expense in the summer of 1687.

Newton made use of geometric methods in all his proofs in the *Principia,* since he realized that his fluxional calculus would be unknown to other mathematicians and might thus lead them to dispute results which were themselves opposed to many of the theories prevalent at the time, such as Descartes' theory of the universe.

Another factor that probably weighed with Newton in reaching his decision to employ the familiar Greek geometry was that the calculus had not been fully developed when he wrote the *Principia,* and consequently was not then as

superior to Greek geometry as it later became. Since Newton gave his geometric demonstrations without explanations,
only outstanding mathematicians were able to follow his
concise reasoning. Nevertheless, thanks to their enthusiastic
acceptance of the book, Newton's theory of the univer:e
soon found widespread acknowledgment, except in France,
where his views met with opposition for many years. In
1736, however, Voltaire, with the aid of his friend, Madame
du Châtelet, a distinguished mathematician, wrote a long
treatise on the Newtonian system, which led to its acceptance
in France as elsewhere in Europe. So great was the demand
for the *Principia* that by 1691 it was impossible to purchase
a copy of the work.

In 1684 Leibnitz made public his methods [for the differential calculus] in a scientific paper he had founded,
called *Acta Eruditorum.* Thus it came about that although
Newton had invented his method of fluxions many years
before Leibnitz had invented his method of "differences,"
Leibnitz was the first of the two to publish his method.

At this time the two great men were on friendly and
cordial terms. From 1684 until 1699 no suggestion was made
that Leibnitz was not the inventor of his own particular
calculus differentialis. Then, in 1699, an obscure Swiss
mathematician, who was living in England and who had
been angered at having been omitted from a list of eminent
mathematicians drawn up by Leibnitz, insinuated in a
paper read before the Royal Society that Leibnitz had not
invented his form of calculus but had based it on Newton's
method of fluxions.

Naturally, Leibnitz was annoyed. He replied to the attack, in the *Acta Eruditorum,* and protested to the Royal
Society.

Nothing more might have been heard of the wretched
business had not Leibnitz in 1705 written an unfavorable
review of the first account of fluxions to be published by
Newton, in 1704, as an appendix to a book dealing with
optics. In this review he remarked that Newton had always
used fluxions instead of his own "differences."

The Savilian Professor of Astronomy at Oxford, John Keill, considered that by this remark Leibnitz had accused Newton of plagiarism, and then proceeded to accuse Leibnitz himself of having published Newton's method as his own, merely changing its name and notation.

Once more, Leibnitz appealed to the Royal Society, of which Newton had been elected president in 1703 (he remained president until his death twenty-five years later). Leibnitz requested that Keill should be induced to withdraw the imputation that he had stolen his method from Newton. A tactful remark from Newton as president might even now have saved the situation. Unfortunately, it was not forthcoming.

The dispute now went from bad to worse and continued long after the death of Leibnitz in 1716 and that of Newton in 1727. In both England and Germany the rights and wrongs of the senseless controversy became submerged in questions of national pride and prestige. Today, after an interval of two centuries, the dispute seems inconsequential and fantastic. Two great minds were at work on the material provided by Kepler, Cavalieri, Fermat, Pascal, Wallis and Barrow. It is not surprising that they both reached similar conclusions.

From the point of view of mathematics, Newton must be regarded as having wasted the last thirty years of his life in occupations unworthy of his supreme talents. While he was living in London, as a Member of Parliament for Cambridge University, he made the acquaintance of John Locke, the great philosopher who was also intimately connected with English politicians, himself being for a time Secretary of the Board of Trade. Locke and other friends of Newton felt that it was outrageous that the most eminent scientist and mathematician of his age should be dependent on the meager salary of a college professor and fellow. By their efforts, Newton was appointed Warden of the Mint in 1695, an appointment worth £500 a year but making so few demands on his time as to allow him to retain his Cambridge professorship. In 1697 he was appointed Master of the Mint, a post worth two to three times as much as that of Warden.

During this period of his life, Newton paid much attention to theological and philosophical studies in addition to the arduous duties he undertook as Master of the Mint.

Toward the end of his long life, he was troubled with illness, though he continued to preside over the Royal Society. He died on March 20, 1727 and was buried in Westminster Abbey.

It is impossible to avoid a feeling of regret that Newton should have allowed himself to be persuaded to degrade his unrivaled talents by accepting a well-paid government post that robbed the world of his genius for a quarter of a century. Had he, like Archimedes, given up all his long life and all his talents to science and mathematics, there is no knowing what further great advances in human knowledge might have been achieved in the quiet seclusion of a Cambridge college.

As it was, in the twenty-two years between 1665 and 1687, he left for all time the stamp of his supreme genius on mathematics and physical science. It is to the author of the *Principia* and the inventor of fluxions and not to the Master of the Mint that the world looks back with gratitude and awe as the first mind in eighteen centuries that equalled the mind of Archimedes.

It is now generally agreed that Newton and Leibnitz invented the calculus simultaneously and independently. The latter was distinguished not only as a scientist and mathematician, but also as a philosopher and diplomat. He invented a calculating machine and was a forerunner of the modern school of symbolic logic. His notation for the calculus was much more easily understood than that of Newton, but the British, for reasons of national pride, refused to accept it, and their mathematics was severely retarded thereby. On the continent, where the method of Leibnitz was accepted, progress was rapid. If Newton had stood on the shoulders of giants, the contributions of

the mathematicians of the eighteenth century were in turn
largely in an understanding and development of the invention
of Leibnitz and Newton.

Using this tool, they made great strides in both pure and
applied mathematics. The Bernoulli brothers, Jacques and Jean,
friends of Leibnitz, popularized his notation. This extraordinary Swiss family produced at least six other mathematicians of
note. Family squabbles, including a dispute between the two
brothers about the theft of a solution of the isoperimetric
problem, were common among them. They were, however, the
most successful mathematics teachers of the age. Among their
students was Jean-le-Rond D'Alembert, an illegitimate child
abandoned by his mother in infancy, who became noted for
his work in dynamics.

As the social ferment which led to the French Revolution
rose to its climax, activity in mathematics paralleled it. Probably the most remarkable mathematician of the period was
Leonhard Euler, born in Basle in 1707. He was a student of
Jean Bernoulli and later joined the younger Bernoulli in St.
Petersburg, where they had been invited by Catherine I. He
lost the use of an eye as a young man, and in Russia became
totally blind. But his prodigious output continued until his
death in 1783. This output was encyclopedic in scope—he
reorganized almost all the branches of mathematics. In addition
he made original contributions, particularly in analysis, but
also in many other fields ranging from mathematical philosophy
to optics. His work on the theory of the moon's motion remains of importance to the present day.

Another giant of the period was J. L. Lagrange. He was born
in Turin in 1736 and had little formal education in mathematics. He became interested in the subject at seventeen,
when he accidentally came across a memoir by Halley on algebra and optics. Two years later he solved a problem in the
calculus of variations which had puzzled mathematicians
for half a century. His Mécanique Analytique, a presentation
of Newtonian mechanics, became a classic. He made advances
in almost every branch of pure mathematics, and his long,
arduous, and original labors earned for him the soubriquet,
"the greatest mathematician in Europe."

Other great names of the period were those of P. S. Laplace, who applied many of Lagrange's discoveries in pure mathematics to physics and astronomy—an example was the Laplace hypothesis of the origin of the solar system; Legendre, the mathematical analyst; and Monge and Poncelet, who founded, respectively, modern descriptive and projective geometry.

With the almost feverish activity of these men and their colleagues, the period just preceding modern mathematics may be said to have come to an end. In the same book from which we have taken the selection on Gauss, Eric Temple Bell describes how Lagrange, wearying of mathematics in his middle age, considered that the subject was exhausted, that "it was finished or at least passing into a period of decadence." In a sense he was right. Had not Gauss, Abel, Galois, Cauchy, and others injected new ideas into mathematics, the surge of the Newtonian impulse would have spent itself. Similar predictions have been made, of course, in other branches of science—almost invariably to be proved mistaken; an example is a leading British physicist's prediction, just before the discovery of radioactivity, that physics had exhausted its potentialities. Actually, mathematics was on the threshold of its brightest age, and the greatest of those who ushered it in was Johann Friedrich Carl Gauss.

The author of our examination of Gauss's contribution to mathematics, Eric Temple Bell, was born in Aberdeen in 1883, and studied mathematics in both England and the United States. In 1926, he became Professor of Mathematics at the California Institute of Technology, a post he held until his retirement shortly before his death in 1960. He was the author of numerous books on mathematics and, surprisingly, of highly competent science-fiction stories, published under a pseudonym. He won many honors for his mathematical research.

THE PRINCE OF MATHEMATICIANS: GAUSS

ERIC TEMPLE BELL

ARCHIMEDES, NEWTON, AND GAUSS, these three, are in a class by themselves among the great mathematicians, and it is not for ordinary mortals to attempt to range them in order of merit.

The lineage of Gauss, Prince of Mathematicians, was anything but royal. The son of poor parents, he was born in a miserable cottage at Brunswick (Braunschweig), Germany, on April 30, 1777. His paternal grandfather was a poor peasant. In 1740 this grandfather settled in Brunswick, where he drudged out a meager existence as a gardener. The second of his three sons, Gerhard Diederich, born in 1744, became the father of Gauss. Beyond that unique honor Gerhard's life of hard labor as a gardener, canal tender, and bricklayer was without distinction of any kind.

The picture we get of Gauss' father is that of an upright, scrupulously honest, uncouth man whose harshness to his sons sometimes bordered on brutality. His speech was rough and his hand heavy. Honesty and persistence gradually won him some measure of comfort, but his circumstances were never easy. It is not surprising that such a man did everything in his power to thwart his young son and prevent him from acquiring an education suited to his abilities. Had the father prevailed, the gifted boy would have followed one of the family trades, and it was only by a series of happy accidents that Gauss was saved from becoming a gardener or a bricklayer.

On his mother's side Gauss was indeed fortunate. Gauss' mother was a forthright woman of strong character, sharp intellect, and humorous good sense. Her son was her pride from the day of his birth to her own death at the age of ninety-seven.

Dorothea hoped and expected great things of her son. That she may sometimes have doubted whether her dreams were to be realized is shown by her hesitant questioning of those in a position to judge her son's abilities. Thus, when Gauss was nineteen, she asked his mathematical friend Wolfgang Bolyai, whether Gauss would ever amount to anything. When Bolyai exclaimed "The greatest mathematician in Europe!" she burst into tears.

In all the history of mathematics there is nothing approaching the precocity of Gauss as a child. It is not known when Archimedes first gave evidence of genius. Newton's earliest manifestations of the highest mathematical talent may well have passed unnoticed. Although it seems incredible, Gauss showed his caliber before he was three years old.

One Saturday Gerhard Gauss was making out the weekly payroll for the laborers under his charge, unaware that his young son was following the proceedings with critical attention. Coming to the end of his long computations, Gerhard was startled to hear the little boy pipe up, "Father, the reckoning is wrong, it should be. . . ." A check of the account showed that the figure named by Gauss was correct.

Shortly after his seventh birthday Gauss entered his first school, a squalid relic of the Middle Ages run by a virile brute, one Büttner.

Nothing extraordinary happened during the first two years. Then, in his tenth year, Gauss was admitted to the class in arithmetic. As it was the beginning class none of the boys had ever heard of an arithmetical progression. It was easy then for the heroic Büttner to give out a long problem in addition whose answer he could find by a formula in a few seconds. The problem was of the following sort, $81297 + 81495 + 81693 + \ldots + 100899$, where the step from one number to the next is the same all along (here 198), and a given number of terms (here 100) are to be added.

It was the custom of the school for the boy who first got the answer to lay his slate on the table; the next laid his slate on top of the first, and so on. Büttner had barely finished stating the problem when Gauss flung his slate on the table: "There it lies," he said—*Ligget se* in his peasant dialect. Then, for the ensuing hour, while the other boys toiled, he sat with his hands folded, favored now and then by a sarcastic glance from Büttner, who imagined the youngest pupil in the class was just another blockhead. At the end of the period Büttner looked over the slates. On Gauss' slate there appeared but a single number. To the end of his days Gauss loved to tell how the one number he had written was the correct answer and how all the others were wrong. Gauss had not been shown the trick for doing such problems rapidly. It is very ordinary once it is known, but for a boy of ten to find it instantaneously by himself is not so ordinary.

This opened the door through which Gauss passed on to immortality. Büttner was so astonished at what the boy of ten had done without instruction that out of his own pocket he paid for the best textbook on arithmetic obtainable and presented it to Gauss. The boy flashed through the book. "He is beyond me," Büttner said; "I can teach him nothing more."

By himself Büttner could probably not have done much for the young genius. But by a lucky chance the schoolmaster had an assistant, Johann Martin Bartels (1769–1836), a young man with a passion for mathematics, whose duty it was to help the beginners in writing and cut their quill pens for them. Between the assistant of seventeen and the pupil of ten there sprang up a warm friendship which lasted out Bartels' life. They studied together, helping one another over difficulties and amplifying the proofs in their common textbook on algebra and the rudiments of analysis.

Bartels did more for Gauss than to induct him into the mysteries of algebra. The young teacher was acquainted with some of the influential men of Brunswick. He now made it his business to interest these men in his find. They in turn, favorably impressed by the obvious genius of Gauss, brought him to the attention of Carl Wilhelm Ferdinand, Duke of Brunswick.

The Duke received Gauss for the first time in 1791. Gauss was then fourteen. The boy's modesty and awkward shyness won the heart of the generous Duke. Gauss left with the assurance that his education would be continued. The following year (February, 1792) Gauss matriculated at the Collegium Carolinum in Brunswick. The Duke paid the bills and he continued to pay them till Gauss' education was finished.

Before entering the Caroline College at the age of fifteen, Gauss had made great headway in the classical languages by private study and help from older friends, thus precipitating a crisis in his career. To his crassly practical father the study of ancient languages was the height of folly. Dorothea Gauss put up a fight for her boy, won, and the Duke subsidized a two-years' course at the Gymnasium. There Gauss' lightning mastery of the classics astonished teachers and students alike. Gauss was strongly attracted to philological studies, but fortunately for science he was presently to find a more compelling attraction in mathematics.

Gauss studied at the Caroline College for three years, during which he mastered the more important works of Euler, Lagrange and, above all, Newton's *Principia.* The highest praise one great man can get is from another in his own class. Gauss never lowered the estimate which as a boy of seventeen he had formed of Newton. Others—Euler, Laplace, Lagrange, Legendre—appear in the flowing Latin of Gauss with the complimentary *clarissimus;* Newton is *summus.*

While still at the college Gauss had begun those researches in the higher arithmetic which were to make him immortal. His prodigious powers of calculation now came into play. Going directly to the numbers themselves he experimented with them, discovering by induction recondite general theorems whose proofs were to cost even him an effort. In this way he rediscovered "the gem of arithmetic," *theorema aureum,* which Euler also had come upon inductively, which is known as the law of quadratic reciprocity, and which he was to be the first to prove.

The whole investigation originated in a simple question which many beginners in arithmetic ask themselves: How many digits are there in the period of a repeating decimal?

To get some light on the problem Gauss calculated the decimal representations of all the fractions $1/n$ for $n = 1$ to 1000. He did not find the treasure he was seeking, but something infinitely greater—the law of quadratic reciprocity.

It was not easy to prove. In fact it baffled Euler and Legendre. Gauss gave the first proof at the age of nineteen. As this reciprocity is of fundamental importance in the higher arithmetic and in many advanced parts of algebra, Gauss turned it over and over in his mind for many years, seeking to find its taproot, until in all he had given six distinct proofs, one of which depends upon the straightedge and compass construction of regular polygons.

When Gauss left the Caroline College in October, 1795 at the age of eighteen to enter the University of Göttingen he was still undecided whether to follow mathematics or philology as his life work. He had already invented (when he was eighteen) the method of "least squares," which today is indispensable in geodetic surveying, in the reduction of observations, and indeed in all work where the "most probable" value of anything that is measured is to be inferred from a large number of measurements. The Gaussian law of normal distribution of errors and its accompanying bell-shaped curve is familiar today to all who handle statistics, from high-minded intelligence testers to unscrupulous market manipulators.

March 30, 1796, marks the turning point in Gauss' career. On that day, exactly a month before his twentieth year opened, Gauss definitely decided in favor of mathematics.

The same day Gauss began to keep his scientific diary (*Notizenjournal*). It is one of the most precious documents in the history of mathematics.

The diary came into scientific circulation only in 1898, forty-three years after the death of Gauss, when the Royal Society of Göttingen borrowed it from a grandson of Gauss for critical study. It consists of nineteen small octavo pages and contains 146 extremely brief statements of discoveries or results of calculations, the last of which is dated July 9, 1814.

Things were buried for years or decades in this diary that would have made half a dozen great reputations had they been published promptly. Some were never made public during Gauss' lifetime, and he never claimed in anything he himself printed to have anticipated others when they caught up with him. But the record stands. These anticipations were not mere trivialities. Some of them became major fields of nineteenth-century mathematics.

One in particular is of the first importance; the entry for March 19, 1797, shows that Gauss had already discovered the double periodicity of certain elliptic functions. He was then not quite twenty. Again, a later entry shows that Gauss had recognized the double periodicity in the general case. This discovery of itself, had he published it, would have made him famous. But he never published it.

Why did Gauss hold back the great things he discovered? This is easier to explain than his genius—if we accept his own simple statements.

Gauss said that he undertook his scientific works only in response to the deepest promptings of his nature, and it was a wholly secondary consideration to him whether they were ever published for the instruction of others. Another statement which Gauss once made to a friend explains both his diary and his slowness in publication. He declared that such an overwhelming horde of new ideas stormed his mind before he was twenty that he could hardly control them and had time to record but a small fraction. The diary contains only the final brief statements of the outcome of elaborate investigations, some of which occupied him for weeks. Contemplating as a youth the close, unbreakable chains of synthetic proofs in which Archimedes and Newton had tamed their inspirations, Gauss resolved to follow their great example and leave after him only finished works of art, severely perfect, to which nothing could be added and from which nothing could be taken away without disfiguring the whole. All traces of the steps by which the goal had been attained having been obliterated, it was not easy for the followers of Gauss to rediscover the road he had traveled. Consequently some of his works had to wait for highly gifted interpreters before mathematicians in general could understand them, see their significance for unsolved problems, and go ahead. His own contemporaries begged him to relax his frigid perfection so that mathematics might advance more rapidly, but Gauss never relaxed. Not till long after his death was it known how much of nineteenth-century mathematics Gauss had foreseen and anticipated before the year 1800. Had he divulged what he knew it is quite possible that mathematics would now be half a century or more ahead of where it is. Abel and Jacobi could have begun where Gauss left off, instead of expending much of their finest effort rediscovering things Gauss knew before they were born, and the creators of non-Euclidean geometry could have turned their genius to other things.

The three years (October, 1795–September, 1798) at the University of Göttingen were the most prolific in Gauss' life. Owing to the generosity of the Duke Ferdinand the young man did not have to worry about finances. He lost himself in his work, making but few friends. One of these, Wolfgang Bolyai, "the rarest spirit I ever knew," as Gauss described him, was to become a friend for life. The course of this friendship and its importance in the history of non-Euclidean geometry is too long to be told here; Wolfgang's son Johann was to retrace practically the same path that Gauss had followed to the creation of a non-Euclidean geometry, in entire ignorance that his father's old friend had anticipated him. The ideas which had overwhelmed Gauss since his seventeenth year were now caught—partly —and reduced to order. Since 1795 he had been meditating a great work on the theory of numbers. This now took definite shape, and by 1798 the *Disquisitiones Arithmeticae* (Arithmetical Researches) was practically completed.

The *Disquisitiones* was the first of Gauss' masterpieces and by some considered his greatest. It was his farewell to pure mathematics as an exclusive interest. After its publication in 1801 (Gauss was then twenty-four), he broadened his activity to include astronomy, geodesy, and electromagnetism in both their mathematical and practical aspects.

But arithmetic was his first love, and he regretted in later life that he had never found the time to write the second volume he had planned as a young man. The book is in seven "sections." There was to have been an eighth, but this was omitted to keep down the cost of printing.

The opening sentence of the preface describes the general scope of the book. "The researches contained in this work appertain to that part of mathematics which is concerned with integral numbers, also fractions, surds [irrationals] being always excluded."

Hampered by the classic perfection of its style the *Disquisitiones* was somewhat slow of assimilation, and when finally gifted young men began studying the work deeply they were unable to purchase copies, owing to the bankruptcy of a bookseller. Even Eisenstein, Gauss' favorite disciple, never owned a copy. Dirichlet was more fortunate. His copy accompanied him on all his travels, and he slept with it under his pillow. Before going to bed he would struggle with some tough paragraph in the hope—frequently fulfilled—that he would wake up in the night to find that a rereading made everything clear.

The second great stage in Gauss' career began on the first day of the nineteenth century, also a red-letter day in the histories of philosophy and astronomy. Since 1781 when Sir William Herschel discovered the planet Uranus, thus bringing the number of planets then known up to the philosophically satisfying seven, astronomers had been diligently searching the heavens for further members of the Sun's family, whose existence was to be expected, according to Bode's law, between the orbits of Mars and Jupiter. The search was fruitless till Giuseppe Piazzi of Palermo, on the opening day of the nineteenth century, observed what he at first mistook for a small comet approaching the Sun, but which was presently recognized as a new planet—later named Ceres, the first of the swarm of minor planets known today.

To compute an orbit from the meager data available was a task which might have exercised Laplace himself. Newton had declared that such problems are among the most difficult in mathematical astronomy. The mere arithmetic necessary to establish an orbit with accuracy sufficient to ensure that Ceres on her whirl round the sun should not be lost to telescopes might well deter an electrically-driven calculating machine even today; but to the young man whose inhuman memory enabled him to dispense with a table of logarithms when he was hard pressed or too lazy to reach for one, all this endless arithmetic—*logistica,* not *arithmetica*—was the sport of an infant.

Ceres was rediscovered, precisely where the marvellously ingenious and detailed calculations of the young Gauss had predicted she must be found. Pallas, Vesta, and Juno, insignificant sister planets of the diminutive Ceres were quickly picked up by prying telescopes, and their orbits, too, were found to conform to the inspired calculations of Gauss. Computations which would have taken Euler three days to perform—one such is sometimes said to have blinded him —were now the simple exercises of a few laborious hours. Gauss had prescribed the *method,* the routine. The major part of his own time for nearly twenty years was devoted to astronomical calculations.

In 1809 he published his second masterpiece, *Theoria motus corporum coelestium in sectionibus conicis solem ambientium* (Theory of the Motion of the Heavenly Bodies Revolving round the Sun in Conic Sections), in which an exhaustive discussion of the determination of planetary and cometary orbits from observational data, including the difficult analysis of perturbations, lays down the law which for many years is to dominate computational and practical astronomy. It was great work, but not as great as Gauss was easily capable of had he developed the hints lying neglected in his diary. No essentially new discovery was added to *mathematics* by the *Theoria motus.*

Recognition came with spectacular promptness after the rediscovery of Ceres. Laplace hailed the young mathematician at once as an equal and presently as a superior. Some time later when the Baron Alexander von Humboldt, the famous traveller and amateur of the sciences, asked Laplace who was the greatest mathematician in Germany, Laplace replied "Pfaff." "But what about Gauss?" the astonished Von Humboldt asked, as he was backing Gauss for the position of director at the Göttingen Observatory. "Oh," said Laplace, "Gauss is the greatest mathematician in the world."

The decade following the Ceres episode was rich in both happiness and sorrow for Gauss. He was not without detractors even at that early stage of his career. Eminent men who had the ear of the polite public ridiculed the young man of twenty-four for wasting his time on so useless a pastime as the computation of a minor planet's orbit. Ceres might be the goddess of the fields, but it was obvious to the merry wits that no corn grown on the new planet would ever find its way into the Brunswick market of a Saturday afternoon. No doubt they were right, but they also ridiculed him in the same way thirty years later when he laid the foundations of the mathematical theory of electromagnetism and invented the electric telegraph. Gauss let them enjoy their jests. He never replied publicly, but in private expressed his regret that men of honor and priests of science could stultify themselves by being so petty. In the meantime he went on with his work, grateful for the honors the learned societies of Europe showered on him but not going out of his way to invite them.

The Duke of Brunswick increased the young man's pension and made it possible for him to marry (October 9, 1805) at the age of twenty-eight. The lady was Johanne Osthof of Brunswick. Writing to his old university friend, Wolfgang Bolyai, three days after he became engaged, Gauss expresses his unbelievable happiness. "Life stands still before me like an eternal spring with new and brilliant colors."

Three children were born of this marriage: Joseph, Minna, and Louis, the first of whom is said to have inherited his father's gift for mental calculations. Johanne died on October 11, 1809, after the birth of Louis, leaving her young husband desolate. His eternal spring turned to winter. Although he married again the following year for the sake of his young children it was long before Gauss could speak without emotion of his first wife. By the second wife, Minna Waldeck, who had been a close friend of the first, he had two sons and a daughter.

In 1808 Gauss lost his father. Two years previously he had suffered an even severer loss in the death of his benefactor under tragic circumstances. His generous patron dead, it became necessary for Gauss to find some reliable livelihood to support his family. There was no difficulty about this as the young mathematician's fame had now spread to the farthest corners of Europe. St. Petersburg had been angling for him as the logical successor of Euler who had never been worthily replaced after his death in 1783. In 1807 a definite and flattering offer was tendered Gauss. Alexander von Humboldt and other influential friends, reluctant to see Germany lose the greatest mathematician in the world, bestirred themselves, and Gauss was appointed director of the Göttingen Observatory with the privilege—and duty, when necessary—of lecturing on mathematics to university students.

The year 1811 might have been a landmark in mathematics comparable to 1801—the year in which the *Disquisitiones Arithmeticae* appeared—had Gauss made public a discovery he confided to Bessel. Having thoroughly understood complex numbers and their geometrical representation as points on the plane of analytic geometry, Gauss proposed himself the problem of investigating what are today called "analytic functions" of such numbers.

Some idea of the importance of analytic functions can be inferred from the fact that vast tracts of the theories of fluid motion (also of mathematical electricity and representation by maps which do not distort angles) are naturally handled by the theory of analytic functions of a complex variable. The theory was one of the greatest fields of mathematical triumphs in the nineteenth century. Gauss in his letter to Bessel states what amounts to the fundamental theorem in this vast theory, but he hid it away to be rediscovered by Cauchy and later Weierstrass.

Equally significant advances in geometry and the applications of mathematics to geodesy, the Newtonian theory of attraction, and electromagnetism were also to be made by Gauss. How was it possible for one man to accomplish this colossal mass of work of the highest order? With characteristic modesty Gauss declared that "If others would but reflect on mathematical truths as deeply and as continuously as I have, they would make my discoveries." Possibly. Gauss' explanation recalls Newton's. Asked how he had made discoveries in astronomy surpassing those of all his predecessors, Newton replied, "By always thinking about them." This may have been plain to Newton; it is not to ordinary mortals.

Part of the riddle of Gauss is answered by his *involuntary* preoccupation with mathematical ideas—which itself of course demands explanation. As a young man Gauss would be "seized" by mathematics. Conversing with friends he would suddenly go silent, overwhelmed by thoughts beyond his control, and stand staring rigidly oblivious of his surroundings. Later he controlled his thoughts—or they lost their control over him—and he consciously directed all his energies to the solution of a difficulty till he succeeded. A problem once grasped was never released till he had conquered it, although several might be in the foreground of his attention simultaneously.

In one such instance (referring to the *Disquisitiones*, page 636) he relates how for four years scarcely a week passed that he did not spend some time trying to settle whether a certain sign should be plus or minus. The solution finally came of itself in a flash. But to imagine that it would have blazed out of itself like a new star without the "wasted" hours is to miss the point entirely. Often after spending days or weeks fruitlessly over some research Gauss would find on resuming work after a sleepless night that the obscurity had vanished and the whole solution shone clear in his mind. The capacity for intense and prolonged concentration was part of his secret.

In this ability to forget himself in the world of his own thoughts Gauss resembles both Archimedes and Newton. In two further respects he also measures up to them, his gifts for precise observation and a scientific inventiveness which enabled him to devise the instruments necessary for his scientific researches. To Gauss geodesy owes the invention of the heliotrope, an ingenious device by which signals could be transmitted practically instantaneously by means of reflected light. For its time the heliotrope was a long step forward. The astronomical instruments he used also received notable improvements at Gauss' hands. For use in his fundamental researches in electromagnetism Gauss invented the bifilar magnetometer. And as a final example of his mechanical ingenuity it may be recalled that Gauss in 1833 invented the electric telegraph and that he and his fellow worker Wilhelm Weber (1804–1891) used it as a matter of course in sending messages. The combination of mathematical genius with first-rate experimental ability is one of the rarest in all science.

Gauss himself cared but little for the possible practical uses of his inventions. Like Archimedes he preferred mathematics to all the kingdoms of the earth; others might gather the tangible fruits of his labors. But Weber, his collaborator in electromagnetic researches, saw clearly what the puny little telegraph of Göttingen meant for civilization.

Unlike Newton in his later years, Gauss was never attracted by the rewards of public office, although his keen interest and sagacity in all matters pertaining to the sciences of statistics, insurance, and "political arithmetic" would have made him a good minister of finance. Till his last illness he found complete satisfaction in his science and his simple recreations. Wide reading in the literature of Europe and the classics of antiquity, a critical interest in world politics, and the mastery of foreign languages and new sciences (including botany and mineralogy) were his hobbies.

For the last twenty-seven years of his life Gauss slept away from his observatory only once, when he attended a scientific meeting in Berlin to please Alexander von Humboldt who wished to show him off.

It would take a long book (possibly a longer one than would be required for Newton) to describe all the outstanding contributions of Gauss to mathematics, both pure and applied. As a rough but convenient table of dates (from that adopted by the editors of Gauss' works) we summarize the principal fields of Gauss' interests after 1800 as follows: 1800–1820, astronomy; 1820–1830, geodesy, the theories of surfaces, and conformal mapping; 1830–1840, mathematical physics, particularly electromagnetism, terrestrial magnetism, and the theory of attraction according to the Newtonian law; 1841–1855, *analysis situs*, and the geometry associated with functions of a complex variable.

During the period 1821–1848 Gauss was scientific adviser to the Hanoverian (Göttingen was then under the government of Hanover) and Danish governments in an extensive geodetic survey. Gauss threw himself into the work. His method of least squares and his skill in devising schemes for handling masses of numerical data had full scope but, more importantly, the problems arising in the precise survey of a portion of the earth's surface undoubtedly suggested deeper and more general problems connected with all curved surfaces.

Geodetic researches also suggested to Gauss the development of another powerful method in geometry, that of conformal mapping. Before a map can be drawn, say of Greenland, it is necessary to determine what is to be preserved. Are distances to be distorted, as they are on Mercator's projection, till Greenland assumes an exaggerated importance in comparison with North America? Or are distances to be preserved, so that one inch on the map, measured anywhere along the reference lines (say those for latitude and longitude) shall always correspond to the same distance measured on the surface of the earth? If so, one kind of mapping is demanded, and this kind will not preserve some other feature that we may wish to preserve; for example, if two roads on the earth intersect at a certain angle, the lines representing these roads on the map will intersect at a different angle. That kind of mapping which preserves angles is called conformal. In such mapping the theory of analytic functions of a complex variable is the most useful tool.

The whole subject of conformal mapping is of constant

use in mathematical physics and its applications, for example in electrostatics, hydrodynamics and its offspring aerodynamics, in the last of which it plays a part in the theory of the airfoil.

Another field of geometry which Gauss cultivated with his usual thoroughness and success was that of the applicability of surfaces, in which it is required to determine what surfaces can be bent onto a given surface without stretching or tearing. Here again the methods Gauss invented were general and of wide utility.

To other departments of science Gauss contributed fundamental researches, for example in the mathematical theories of electromagnetism, including terrestrial magnetism, capillarity, the attraction of ellipsoids (the planets are special kinds of ellipsoids) where the law of attraction is the Newtonian, and dioptrics, especially concerning systems of lenses. The last gave him an opportunity to apply some of the purely abstract technique (continued fractions) he had developed as a young man to satisfy his curiosity in the theory of numbers.

Gauss not only mathematicized sublimely about all these things; he used his hands and his eyes, and was an extremely accurate observer. Many of the specific theorems he discovered, particularly in his researches on electromagnetism and the theory of attraction, have become part of the indispensable stock in trade of all who work seriously in physical science. For many years Gauss, aided by his friend Weber, sought a satisfying theory for all electromagnetic phenomena. Failing to find one that he considered satisfactory he abandoned his attempt. Had he found Clerk Maxwell's (1831–1879) equations of the electromagnetic field he might have been satisfied.

To conclude this long but still far from complete list of the great things that earned Gauss the undisputed title of Prince of Mathematicians we must allude to a subject on which he published nothing beyond a passing mention in his dissertation of 1799, but which he predicted would become one of the chief concerns of mathematics—*analysis situs*. A technical definition of what this means is impossible here (it requires the notion of a continuous group), but some hint of the type of problem with which the subject deals can be gathered from a simple instance. Any sort of a knot is tied in a string, and the ends of the string are then tied together. A "simple" knot is easily distinguishable by eye from a "complicated" one, but how are we to give an exact, mathematical specification of the difference between the two? And how are we to classify knots mathematically? Although he published nothing on this, Gauss had made a beginning, as was discovered in his posthumous papers. Another type of problem in this subject is to determine the least number of cuts on a given surface which will enable us to flatten the surface out on a plane. For a conical surface one cut suffices; for an anchor ring, two; for a sphere, no finite number of cuts suffices if no stretching is permitted.

These examples may suggest that the whole subject is trivial. But if it had been, Gauss would not have attached the extraordinary importance to it that he did. His prediction of its fundamental character has been fulfilled in our own generation. Today a vigorous school is finding that *analysis situs,* or the "geometry of position" as it used sometimes to be called, has far-reaching ramifications in both geometry and analysis.

His last years were full of honor, but he was not as happy as he had earned the right to be. As powerful of mind and as prolifically inventive as he had ever been, Gauss was not eager for rest when the first symptoms of his last illness appeared some months before his death.

A narrow escape from a violent death had made him more reserved than ever, and he could not bring himself to speak of the sudden passing of a friend. For the first time in more than twenty years he had left Göttingen on June 16, 1854, to see the railway under construction between his town and Cassel. Gauss had always taken a keen interest in the construction and operation of railroads; now he would see one being built. The horses bolted; he was thrown from his carriage, unhurt, but badly shocked. He recovered, and had the pleasure of witnessing the opening ceremonies when the railway reached Göttingen on July 31, 1854. It was his last day of comfort.

With the opening of the new year he began to suffer greatly from an enlarged heart and shortness of breath, and symptoms of dropsy appeared. Nevertheless he worked when he could, although his hand cramped and his beautifully clear writing broke at last. The last letter he wrote was to Sir David Brewster on the discovery of the electric telegraph.

Fully conscious almost to the end he died peacefully, after a severe struggle to live, early on the morning of February 23, 1855, in his seventy-eighth year. He lives everywhere in mathematics.

Of the individuals whom we may call the founders of modern mathematics, two—Galois and Abel—led short and tragic lives. Niels Hendrick Abel was born in Norway in 1802, and was left the sole support of his family at 18. An able teacher introduced him to the joys and rigors of mathematics. His first important paper, on the quintic equation, was ignored by contemporary mathematicians—Gauss is said to have tossed it aside in disgust. He was, however, able to obtain a fellowship for foreign study. In France, he offered his great paper on the integrals of algebraic functions to the French Academy. It was a work of monumental importance, but Cauchy, another great mathematician of the period, who was asked to judge its merits, forgot to do so. The paper was ignored. Abel returned home without funds, and managed to sustain a poverty-stricken existence by tutoring. He had contracted tuberculosis and died of the disease in his twenty-sixth year. With classic irony, word came of his appointment as Professor of Mathematics at the University of Berlin two days after his death.

The life of Galois was a similar monument of misunderstanding. With the same clarity and authority which he displayed in his life of Archimedes, George Sarton writes of the

young man who, in attempting to describe his own position,
said, "Genius is condemned by a malicious social organization
to an eternal denial of justice in favor of fawning mediocrity."

EVARISTE GALOIS

GEORGE SARTON

NO EPISODE IN THE HISTORY OF THOUGHT is more moving
than the life of Evariste Galois—the young Frenchman who
passed like a meteor about 1828, devoted a few feverish
years to the most intense meditation, and died in 1832 from
a wound received in a duel, at the age of twenty. He was
still a mere boy, yet within these short years he had accomplished enough to prove indubitably that he was one of the
greatest mathematicians of all time. When one sees how
terribly fast this ardent soul, this wretched and tormented
heart, were consumed, one can but think of the beautiful
meteoric showers of a summer night. But this comparison
is misleading, for the soul of Galois will burn on throughout
the ages and be a perpetual flame of inspiration. His fame
is incorruptible; indeed the apotheosis will become more and
more splendid with the gradual increase of human knowledge.

It is safe to predict that Galois' fame can but wax, because of the fundamental nature of his work. While the
inventors of important applications, whose practical value
is obvious, receive quick recognition and often very substantial rewards, the discoverers of fundamental principles
are not generally awarded much recompense. They often
die misunderstood and unrewarded. But while the fame of
the former is bound to wane as new processes supersede
their own, the fame of the latter can but increase. Indeed
the importance of each principle grows with the number
and the value of its applications; for each new application

is a new tribute to its worth. To put it more concretely, when we are very thirsty a juicy orange is more precious to us than an orange tree. Yet when the emergency has passed, we learn to value the tree more than any of its fruits; for each orange is an end in itself, while the tree represents the innumerable oranges of the future. The fame of Galois has a similar foundation; it is based upon the unlimited future. He well knew the pregnancy of his thoughts, yet they were even more far-reaching than he could possibly dream of. His complete works fill only sixty-one small pages: but a French geometer, publishing a large volume some forty years after Galois' death, declared that it was simply a commentary on the latter's discoveries. Since then, many more consequences have been deduced from Galois' fundamental ideas which have influenced the whole of mathematical philosophy. It is likely that when mathematicians of the future contemplate his personality at the distance of a few centuries, it will appear to them to be surrounded by the same halo of wonder as those of Euclid, Archimedes, Descartes, and Newton.

Evariste Galois was born in Bourg-la-Reine, near Paris, on the 25th of October, 1811, in the very house in which his grandfather had lived and had founded a boys' school. This being one of the very few boarding schools not in the hands of the priests, the Revolution had much increased its prosperity. In the course of time, grandfather Galois had given it up to his younger son and soon after, the school had received from the imperial government a sort of official recognition. When Evariste was born, his father was thirty-six years of age. He had remained a real man of the eighteenth century, amiable and witty, clever at rhyming verses and writing playlets, and instinct with philosophy. He was the leader of liberalism in Bourg-la-Reine, and during the Hundred Days had been appointed its mayor. Strangely enough, after Waterloo he was still the mayor of the village. He took his oath to the King, and to be sure he kept it, yet he remained a liberal to the end of his days. One of his friends and neighbours, Thomas François Demante, a lawyer and judge, onetime professor in the

Faculty of Law of Paris, was also a typical gentleman of
the "ancien régime," but of a different style. He had given
a very solid classical education not only to his sons but also
to his daughters. None of these had been more deeply
imbued with the examples of antiquity than Adelaïde-
Marie who was to be Evariste's mother. Roman stoicism
had sunk deep into her heart and given to it a virile temper.
She was a good Christian, though more concerned with the
ethical than with the mystical side of religion. An ardent
imagination had colored her every virtue with passion.
Many more people have been able to appreciate her character than her son's, for it was to be her sad fortune to
survive him forty years. She was said to be generous to a
fault and original to the point of queerness. There is no
doubt that Evariste owed considerably more to her than
to his father. Besides, until the age of eleven the little boy
had no teacher but his mother.

In 1823, Evariste was sent to college in Paris. This college
—Louis-le-Grand—was then a gloomy house, looking from
the oustide like a prison, but within aflame with life and
passion. For heroic memories of the Revolution and the
Empire had remained particularly vivid in this institution,
which was indeed, under the clerical and reactionary
regime of the Restoration, a hot-bed of liberalism. Love
of learning and contempt of the Bourbons divided the hearts
of the scholars. Since 1815 the discipline had been jeopard-
ized over and over again by boyish rebellions, and Evariste
was certainly a witness of, if not a partner in, those which
took place soon after his arrival. The influence of such an
impassioned atmosphere upon a lad freshly emancipated
from his mother's care cannot be exaggerated. Nothing is
more infectious than political passion, nothing more in-
toxicating than the love of freedom. It was certainly there
and then that Evariste received his political initiation. It was
the first crisis of his childhood.

At first he was a good student; it was only after a couple
of years that his disgust at the regular studies became
apparent. He was then in the second class (that is, the

highest but one) and the headmaster suggested to his father that he should spend a second year in it, arguing that the boy's weak health and immaturity made it imperative. The child was not strong, but the headmaster had failed to discover the true source of his lassitude. His seeming indifference was due less to immaturity than to his mathematical precocity. He had read his books of geometry as easily as a novel, and the knowledge had remained firmly anchored in his mind. No sooner had he begun to study algebra than he read Lagrange's original memoirs. This extraordinary facility had been at first a revelation to himself, but as he grew more conscious of it, it became more difficult for him to curb his own domineering thought and to sacrifice it to the routine of classwork. The rigid program of the college was to him like a bed of Procrustes, causing him unbearable torture without adequate compensation. But how could the headmaster and the teachers understand this? The double conflict within the child's mind and between the teachers and himself, as the knowledge of his power increased, was intensely dramatic. By 1827 it had reached a critical point. This might be called the second crisis of his childhood: his scientific initiation. His change of mood was observed by the family. Juvenile gaiety was suddenly replaced by concentration; his manners became stranger every day. A mad desire to march forward along the solitary path which he saw so distinctly, possessed him. His whole being, his every faculty was mobilized in this immense endeavor.

I cannot give a more vivid idea of the growing strife between this inspired boy and his uninspired teachers than by quoting a few extracts from the school reports:

1826–1827

This pupil, though a litte queer in his manners, is very gentle and seems filled with innocence and good qualities. . . . He never knows a lesson badly: either he has not learned it at all or he knows it well. . . .

A little later:

> This pupil, except for the last fortnight during which
> he has worked a little, has done his classwork only from
> fear of punishment. . . . His ambition, his originality —
> often affected—the queerness of his character keep him
> aloof from his companions.

1827–1828

> Conduct rather good. A few thoughtless acts. Character
> of which I do not flatter myself I understand every
> trait; but I see a great deal of self-esteem dominating. I
> regard as much to literary studies as to mathematics. . . .
> He does not seem to lack religious feeling. His health is
> good but delicate.

Another professor says:

> His facility, in which one is supposed to believe but of
> do not think he has any vicious inclination. His ability
> seems to me to be entirely beyond the average, with
> which I have not yet witnessed a single proof, will lead
> him nowhere; there is no trace in his tasks of anything
> but of queerness and negligence.

Another still:

> Always busy with things which are not his business.
> Goes down every day.

Same year, but a little later:

> Very bad conduct. Character rather secretive. Tries to
> be original. . . .
> Does absolutely nothing for the class. The furor of
> mathematics possesses him. . . . He is losing his time
> here and does nothing but torment his masters and get
> himself harassed with punishments. He does not lack
> religious feeling; his health seems weak.

Later still:

> Bad conduct, character difficult to define. Aims at
> originality. His talents are very distinguished; he might
> have done very well in "Rhétorique" if he had been
> willing to work, but swayed by his passion for mathe-
> matics, he has entirely neglected everything else. Hence
> he has made no progress whatever. . . . Seems to affect
> to do something different from what he should do. It is
> possibly to this purpose that he chatters so much. He
> protests against silence.

In his last year at the college, 1828–1829, he had at last
found a teacher of mathematics who divined his genius and
tried to encourage and to help him. This Mr. Richard, to
whom one cannot be too grateful, wrote of him: "This
student has a marked superiority over all his schoolmates.
He works only at the highest parts of mathematics." You see
the whole difference. Kind Mr. Richard did not complain
that Evariste neglected his regular tasks, and, I imagine,
often forgot to do the petty mathematical exercises which
are indispensable to drill the average boy; he does not
think it fair to insist on what Evariste does not do, but
states what he does do: he is only concerned with the
highest parts of mathematics. Unfortunately, the other
teachers were less indulgent. For physics and chemistry,
the note often repeated was: "Very absent-minded, no work
whatever."

To show the sort of preoccupations which engrossed his
mind: at the age of sixteen he believed that he had found
a method of solving general equations of the fifth degree.
One knows that before succeeding in proving the impossibility of such resolution, Abel had made the same mistake.
Besides, Galois was already trying to realize the great dream
of his boyhood: to enter the École Polytechnique. He was
bold enough to prepare himself alone for the entrance
examination as early as 1828—but failed. This failure
was very bitter to him—the more so that he considered it
as unfair. It is likely that it was not at all unfair, at least
according to the accepted rules. Galois knew at one and the

same time far more and far less than was necessary to
enter Polytechnique; his extra knowledge could not com-
pensate for his deficiencies, and examiners will never consider originality with favor. The next year he published his
first paper, and sent his first communication to the Académie
des Sciences. Unfortunately, the latter got lost through
Cauchy's negligence. This embittered Galois even more. A
second failure to enter Polytechnique seemed to be the
climax of his misfortune, but a greater disaster was still in
store for him. On July 2 of this same year, 1829, his father
had been driven to commit suicide by the vicious attacks
directed against him, the liberal mayor, by his political
enemies. He took his life in the small apartment which he
had in Paris, in the vicinity of Louis-le-Grand. As soon as
his father's body reached the territory of Bourg-la-Reine,
the inhabitants carried it on their shoulders, and the funeral
was the occasion of disturbances in the village. This terrible
blow, following many smaller miseries, left a very deep
mark on Evariste's soul. His hatred of injustice became
the more violent, in that he already believed himself to be
a victim of it; his father's death incensed him, and developed
his tendency to see injustice and baseness everywhere.

His repeated failures to be admitted to Polytechnique
were to Galois a cause of intense disappointment. To appreciate his despair, one must realize that the École Polytechnique was then, not simply the highest mathematical
school in France and the place where his genius would be
most likely to find the sympathy it craved, it was also a
daughter of the Revolution who had remained faithful to
her origins in spite of all efforts of the government to curb
her spirit of independence. The young Polytechnicians
were the natural leaders of every political rebellion; liberalism was for them a matter of traditional duty. This
house was thus twice sacred to Galois, and his failure to be
accepted was a double misfortune. In 1829 he entered the
École Normale, but he entered it as an exile from Polytechnique. It was all the more difficult for him to forget
the object of his former ambition, because the École Normale was then passing through the most languid period of

its existence. It was not even an independent institution, but rather an extension of Louis-le-Grand. Every precaution had been taken to ensure the loyalty of this school to the new regime. Yet there, too, the main student body inclined toward liberalism, though their convictions were very weak and passive as compared with the mood prevailing at Polytechnique; because of the discipline and the spying methods to which they were submitted, their aspirations had taken a more subdued and hypocritical form only relieved once in a while by spasmodic upheavals. Evariste suffered doubly, for his political desires were checked and his mathematical ability remained unrecognized. Indeed he was easily embarrassed at the blackboard, and made a poor impression upon his teachers. It is quite possible that he did not try in the least to improve this impression. His French biographer, P. Dupuy, very clearly explains his attitude:

There was in him a hardly disguised contempt for whosoever did not bow spontaneously and immediately before his superiority, a rebellion against a judgment which his conscience challenged beforehand and a sort of unhealthy pleasure in leading it further astray and in turning it entirely against himself. Indeed, it is frequently observed that those people who believe that they have most to complain of persecution could hardly do without it and, if need be, will provoke it. To pass oneself off for a fool is another way and not the least savory, of making fools of others.

It is clear that Galois' temper was not altogether amiable, yet we should not judge him without making full allowance for the terrible strain to which he was constantly submitted, the violent conflicts which obscured his soul, the frightful solitude to which fate had condemned him.

In the course of the ensuing year, he sent three more papers to mathematical journals and a new memoir to the Académie. The permanent secretary, Fourier, took it home with him, but died before having examined it, and the memoir was not retrieved from among his papers. Thus his second memoir was lost like the former. This was too much indeed and one will easily forgive the wretched boy if in his feverish mood he was inclined to believe that these repeated losses were not due to chance but to systematic persecution. He considered himself a victim of a bad social organization which ever sacrifices genius to mediocrity, and naturally enough he cursed the hated regime of oppression which had precipitated his father's death and against which the storm was gathering. We can well imagine his joy when he heard the first shots of the July Revolution! But alas! While the boys of Polytechnique were the very first in the fray, those of the École Normale were kept under lock and key by their faint-hearted director. It was only when the three glorious days of July were over and the fall of the Bourbons was accomplished that this opportunist let his students out and indeed placed them at the disposal of the provisional government! Never did Galois feel more bitterly that his life had been utterly spoiled by his failure to become an alumnus of his beloved Polytechnique.

In the meanwhile the summer holidays began and we do not know what happened to the boy in the interval. It must have been to him a new period of crisis, more acute than any of the previous ones. But before speaking of it let me say a last word about his scientific efforts, for it is probable that thereafter political passion obsessed his mind almost exclusively. At any rate it is certain that Evariste was in the possession of his general principles by the beginning of 1830, that is, at the age of eighteen, and that he fully knew their importance. The consciousness of his power and of the responsibility resulting from it increased the concentration and the gloominess of his mind to the danger point; the lack of recognition developed in him an excessive pride. By a strange aberration he did not trouble himself to write his memoirs with sufficient clearness to give the explanations which were the more necessary because his thoughts were more novel. What a pity that there was no understanding friend to whisper in his ear Descartes' wise admonition: "When you have to deal with transcendent questions, you must be transcendently clear." Instead of that, Galois enveloped his thought in additional secrecy by his efforts to attain greater conciseness, that coquetry of mathematicians.

It is intensely tragic that this boy already sufficiently harassed by the turmoil of his own thoughts, should have been thrown into the political turmoil of this revolutionary period. Endowed with a stronger constitution, he might have been able to cope with one such; but with the two, how could he—how could anyone do it? During the holidays he was probably pressed by his friend, Chevalier, to join the Saint-Simonists, but he declined, and preferred to join a secret society, less aristocratic and more in keeping with his republican aspirations—the "Société des amis du peuple." It was thus quite another man who reentered the École Normale in the autumn of 1830. The great events of which he had been a witness had given to his mind a sort of artificial maturity. The revolution had opened to him a fresh source of disillusion, the deeper because the hopes of the first moment had been so sanguine. The government of Louis-Philippe had promptly crushed the more liberal tendencies; and the artisans of the new revolution, who had drawn their inspiration from the great events of 1789, soon discovered to their intense disgust that they had been fooled. Indeed under a more liberal guise, the same oppression, the same favoritism, the same corruption soon took place under Louis-Philippe as under Charles X. Moreover, nothing can be more demoralizing than a successful revolution (whatever it be) for those who, like Galois, were too generous to seek any personal advantage and too ingenuous not to believe implicitly in their party shibboleths. It is such a high fall from one's dearest ideal to the ugliest aspect of reality—and they could not help seeing around them the more practical and cynical revolutionaries eager for the quarry, and more disgusting still, the clever ones, who had kept quiet until they knew which side was gaining, and who now came out of their hiding places to fight over the spoils and make the most of the new regime. Political fermentation did not abate and the more democratic elements, which Galois had joined, became more and more disaffected and restless. The director of the École Normale had been obliged to restrain himself considerably to brook Galois' irregular conduct, his "laziness," his intractable temper; the boy's political attitude, and chiefly his undisguised contempt for the director's pusillanimity now increased the tension between them to the breaking point. The publication in the "Gazette des Écoles" of a letter of Galois' in which he scornfully criticized the director's tergiversations was but the last of many offenses. On December 9, he was invited to leave the school, and his expulsion was ratified by the Royal Council on January 3, 1831.

To support himself Galois announced that he would give a private course on higher algebra in the back shop of a bookseller, Mr. Caillot, 5 rue de la Sorbonne. I do not know whether this course, or how much of it, was actually delivered. A further scientific disappointment was reserved for him: a new copy of his second lost memoir had been communicated by him to the Académie; it was returned to him by Poisson, four months later, as being incomprehensible. There is no doubt that Galois was partly responsible for this, for he had taken no pains to explain himself clearly.

This was the last straw! Galois' academic career was entirely compromised, the bridges were burned, he plunged himself entirely into the political turmoil. He threw himself into it with his habitual fury and the characteristic intransigency of a mathematician; there was nothing left to conciliate him, no means to moderate his passion, and he soon reached the extreme limit of exaltation. He is said to have exclaimed: "If a corpse were needed to stir the people up, I would give mine." Thus on May 9, 1831, at the end of a political banquet, being intoxicated—not with wine but with the ardent conversation of an evening—he proposed a sarcastic toast to the King. He held his glass and an open knife in one hand and said simply: "To Louis-Philippe!" Of course he was soon arrested and sent to Ste. Pélagie. The lawyer persuaded him to maintain that he had actually said: "To Louis-Philippe, *if he betray,*" and many witnesses affirmed that they had heard him utter the last words, though they were lost in the uproar. But Galois could not stand this lying and retracted it at the public trial. His attitude before the tribunal was ironical and provoking, yet the jury rendered a verdict of not proven and he was acquitted. He did not remain free very long. On the following Fourteenth of July, the government, fearing manifestations, decided to have him arrested as a preventive measure. He was given six months' imprisonment on the technical charge of carrying arms and wearing a military uniform, but he remained in Ste. Pélagie only until March 19 (or 16?), 1832, when he was sent to a convalescent home in the rue de Lourcine. A dreadful epidemic of cholera was then raging in Paris, and Galois' transfer had been determined by the poor state of his health. However, this proved to be his undoing.

He was now a prisoner on parole and took advantage of it to carry on an intrigue with a woman of whom we know nothing, but who was probably not very reputable ("une coquette de bas étage," says Raspail). Think of it! This was, as far as we know, his first love—and it was but one more tragedy on top of so many others. The poor boy who had declared in prison that he could love only a Cornelia or a Tarpeia* (we hear in this an echo of his mother's Roman ideal), gave himself to this new passion with his frenzy, only to find more bitterness at the end of it. His revulsion is lamentably expressed in a letter to Chevalier (May 25, 1832):

. . . How to console oneself for having exhausted in one month the greatest source of happiness which is in man—of having exhausted it without happiness, without hope, being certain that one has drained it for life?

Oh! come and preach peace after that! Come and ask men who suffer to take pity upon what is! Pity, never! Hatred, that is all. He who does not feel it deeply, this hatred of the present, cannot really have in him the love of the future. . . .

One sees how his particular misery and his political grievances are sadly muddled in his tired head. And a little

*He must have quoted Tarpeia by mistake.

further in the same letter, in answer to a gentle warning
by his friend:

> I like to doubt your cruel prophecy when you say
> that I shall not work any more. But I admit it is not
> without likelihood. To be a savant, I should need to be
> that alone. *My heart has revolted against my head.** I do
> not add as you do: It is a pity.

Can a more tragic confession be imagined? One realizes
that there is no question here of a man possessing genius,
but of genius possessing a man. A man? a mere boy, a fragile
little body divided within itself by disproportionate forces,
an undeveloped mind crushed mercilessly between the
exaltation of scientific discovery and the exaltation of senti-
ment.

Four days later two men challenged him to a duel! The
circumstances of this affair are, and will ever remain, very
mysterious. According to Evariste's younger brother the duel
was not fair. Evariste, weak as he was, had to deal with
two ruffians hired to murder him. I find nothing to counte-
nance this theory except that he was challenged by two men
at once. At any rate, it is certain that the woman he had
loved played a part in this fateful event. On the day preceding the duel, Evariste wrote three letters of which I
translate one:

May 29, 1832

LETTER TO ALL REPUBLICANS.

I beg the patriots, my friends, not to reproach me for
dying otherwise than for the country.

I die the victim of an infamous coquette. My life is
quenched in a miserable piece of gossip.

Oh! why do I have to die for such a little thing, to die
for something so contemptible!

I take heaven to witness that it is only under compul-
sion that I have yielded to a provocation which I had
tried to avert by all means.

*The italics are mine.

I repent having told a baleful truth to men who were so little able to listen to it coolly. Yet I have told the truth. I take with me to the grave a conscience free from lie, free from patriots' blood.

Good-bye! I had in me a great deal of life for the public good.

Forgiveness for those who killed me; they are of good faith.

E. Galois

Any comment could but detract from the pathos of this document. I will only remark that the last line, in which Galois absolves his adversaries, destroys his brother's theory. It is simpler to admit that his impetuosity, aggravated by female intrigue, had placed him in an impossible position from which there was no honorable issue, according to the standards of the time, but a duel. Evariste was too much of a gentleman to try to evade the issue, however trifling its causes might be; he was anxious to pay the full price of his folly. That he well realized the tragedy of his life is quite clear from the laconic post-scriptum of his second letter: *Nitens lux, horrenda procella, tenebris æternis involuta.* The last letter addressed to his friend, Auguste Chevalier, was a sort of scientific testament. Its seven pages, hastily written, dated at both ends, contain a summary of the discoveries which he had been unable to develop. This statement is so concise and so full that its significance could be understood only gradually as the theories outlined by him were unfolded by others. It proves the depth of his insight, for it anticipates discoveries of a much later date. At the end of the letter, after requesting his friend to publish it and to ask Jacobi or Gauss to pronounce upon it, he added: "After that, I hope some people will find it profitable to unravel this mess. *Je t'embrasse avec effusion.*"— The first sentence is rather scornful but not untrue and the greatest mathematicians of the century have found it very profitable indeed to clear up Galois' ideas.

The duel took place on the 30th in the early morning, and he was grievously wounded by a shot in the abdomen. He was found by a peasant who transported him at 9:30 to the Hôpital Cochin. His younger brother—the only member of the family to be notified—came and stayed with him, and as he was crying, Evariste tried to console him, saying: "Do not cry. I need all my courage to die at twenty." While still fully conscious, he refused the assistance of a priest. In the evening peritonitis declared itself and he breathed his last at ten o'clock on the following morning.

His funeral, which strangely recalled that of his father, was attended by two to three thousand republicans, including deputations from various schools, and by a large number of police, for trouble was expected. But everything went off very calmly. Of course it was the patriot and the lover of freedom whom all these people meant to honor; little did they know that a day would come when this young political hero would be hailed as one of the greatest mathematicians of all time.

A life as short yet as full as the life of Galois is interesting not simply in itself but even more perhaps because of the light it throws upon the nature of genius. When a great work is the natural culmination of a long existence devoted to one persistent endeavor, it is sometimes difficult to say whether it is the fruit of genius or the fruit of patience. When genius evolves slowly it may be hard to distinguish from talent—but when it explodes suddenly, at the beginning and not at the end of life, or when we are at a loss to explain its intellectual genesis, we can but feel that we are in the sacred presence of something vastly superior to talent. When one is confronted with facts which cannot be explained in the ordinary way, is it not more scientific to admit our ignorance than to hide it behind faked explanations? Of course it is not necessary to introduce any mystical idea, but it is one's duty to acknowledge the mystery. When a work is really the fruit of genius, we cannot conceive that a man of talent might have done it "just as well" by taking the necessary pains. Pains alone will never do; neither is it simply a matter of jumping a little further, for it involves a synthetic process of a higher kind. I do not say that talent and genius are essentially different, but that they are of different orders of magnitude.

Galois' fateful existence helps one to understand Lowell's saying: "Talent is that which is in a man's power, genius is that in whose power man is." If Galois had been simply a mathematician of considerable ability, his life would have been far less tragic, for he could have used his mathematical talent for his own advancement and happiness; instead of which, the furor of mathematics—as one of his teachers said—possessed him and he had no alternative but absolute surrender to his destiny.

Lowell's aphorism is misleading, however, for it suggests that talent can be acquired, while genius cannot. But biological knowledge points to the conclusion that neither is really acquired, though both can be developed and to a certain extent corrected by education. Men of talent as well as men of genius are born, not made. Genius implies a much stronger force, less adaptable to environment, less tractable by education, and also far more exclusive and despotic. Its very intensity explains its frequent precocity. If the necessary opportunities do not arise, ordinary abilities may remain hidden indefinitely; but the stronger the abilities the smaller need the inducement be to awaken them. In the extreme case, the case of genius, the ability is so strong that, if need be, it will force its own outlet.

Thus it is that many of the greatest accomplishments of science, art and letters were conceived by very young men. In the field of mathematics, this precocity is particularly obvious. To speak only of the two men considered in this essay, Abel had barely reached the age of twenty-two and Galois was not yet twenty, perhaps not yet nineteen, when they made two of the most profound discoveries which have ever been made. In many other sciences and arts, technical apprenticeship may be too long to make such early discovery possible. In most cases, however, the judgment of Alfred de Vigny holds good. "What is a great life? It is a thought of youth wrought out in ripening years." The fundamental conception dawns at an early age—that is, it appears at the surface of one's consciousness as early as this is materially possible—but it is often so great that a long life of toil and abnegation is but too short to work it out. Of course at the beginning it may be very vague, so vague indeed that its host can hardly distinguish it himself from a passing fancy, and later may be unable to explain how it gradually took control of his activities and dominated his whole being. The cases of Abel and Galois are not essentially different from those contemplated by Alfred de Vigny, but the golden thoughts of their youth were wrought out in the ripening years of other people.

It is the precocity of genius which makes it so dramatic. When it takes an explosive form, as in the case of Galois, the frail carcass of a boy may be unable to resist the internal strain and it may be positively wrecked. On the other hand when genius develops more slowly, its host has time to mature, to adapt himself to his environment, to gather strength and experience. He learns to reconcile himself to the conditions which surround him, widely different as they are, from those of his dreams. He learns by and by that the great majority of men are rather unintelligent, uneducated, uninspired, and that one must not take it too much to heart when they behave in defiance of justice or even of common sense. He also learns to dissipate his vexation with a smile or a joke and to protect himself under a heavy cloak of kindness and humor. Poor Evariste had no time to learn all this. While his genius grew in him out of all proportion to his bodily strength, his experience, and his wisdom, he felt more and more ill at ease. His increasing restlessness makes one think of that exhibited by people who are prey to a larvate form of a pernicious disease. There is an internal disharmony in both cases, though it is physiological in the latter, and psychological in the former. Hence the suffering, the distress, and finally the acute disease or the revolt!

A more congenial environment might have saved Galois. Oh! would that he had been granted that minimum of understanding and sympathy which the most concentrated mind needs as much as a plant needs the sun! . . . But it was not to be; and not only had he no one to share his own burden, but he had also to bear the anxieties of a stormy time. I quite realize that this self-centered boy was not attractive—many would say not lovable. Yet I love him; I love him for all those who failed to love him; I love him because of his adversity.

His tragic life teaches us at least one great lesson: one can never be too kind to the young; one can never be too tolerant of their faults, even of their intolerance. The pride and intolerance of youth, however immoderate, are excusable because of youth's ignorance, and also because one may hope that it is only a temporary disorder. Of course there will always be men despicable enough to resort to snubbing, as it were, to protect their own position and to hide their mediocrity, but I am not thinking of them. I am simply thinking of the many men who were unkind to Galois without meaning to be so. To be sure, one could hardly expect them to divine the presence of genius in an awkward boy. But even if they did not believe in him, could they not have shown more forbearance? Even if he had been a conceited dunce, instead of a genius, could kindness have harmed him? . . . It is painful to think that a few rays of generosity from the heart of his elders might have saved this boy or at least might have sweetened his life.

But does it really matter? A few years more or less, a little more or less suffering. . . . Life is such a short drive altogether. Galois has accomplished his task and very few men will ever accomplish more. He has conquered the purest kind of immortality. As he wrote to his friends: "I take with me to the grave a conscience free from lie, free from patriots' blood." How many of the conventional heroes of history, how many of the kings, captains and statesmen could say the same?

Practically every historian of mathematics has pointed out the impossibility of preparing a concise and connected history of modern mathematics. Florian Cajori estimated that in order to write the history of nineteenth-century mathematics alone on a detailed level, fourteen or fifteen large volumes would be required. George Sarton considered such an extensive examination unrewarding but he too pointed out the difficulties involved, not only because of the richness and complexity of the subject, but also because of its technical nature—how describe advances whose very nature are impossible of comprehension by the lay reader? The historian of mathematics is driven to discuss it in general cultural terms (a method which Sarton approved) or to select only one branch of the endlessly complicated mathematical tree. In view of Cajori's estimate, it is obvious that whole subjects can barely be mentioned in a volume of this size, while most of them must be omitted altogether.

Ernest W. Brown, late Professor of Astronomy at Yale, attempted to examine mathematics in the nineteenth and twentieth centuries from two points of views "one, the development of the three great branches of mathematics, geometry, analysis and their applications to other studies; the other, the development of new ideas which have applications in many branches of mathematics." The development in geometry which has achieved most attention is the creation of systems in which the famous fifth postulate of Euclid—the so-called parallel postulate—is omitted. Gauss first realized that a consistent geometry could exist without this postulate—but perhaps out of fear of ridicule did not pursue the idea. Lobachevsky, the Russian, and Bolyai, the Hungarian, postulated a system in which through a point there are an infinity of parallels to a given line. Riemann, the German, discovered elliptic geometry, in which the sum of the angles of a triangle is greater than 180° and straight lines are finite in length. All the axioms have been omitted or modified in various systems. The name of the German David Hilbert is outstanding in this connection. Modern physical and astronomical theories envisage space which is non-Euclidean and four-dimensional in character.

Another branch of geometry is analysis situs or topology,

which is examined in a selection on page 208. Among the "purest" of pure branches of mathematics is number theory, in which such problems as finding the number of primes under a certain limit or that of proving that every even number is the sum of two primes, are considered.

The development of the theory of functions received a strong impetus from the works of Cauchy. The subject is commonly divided into the theory of functions of a real and of a complex variable. The latter theory was developed by Weierstrass and Riemann and has been carried on by contemporary mathematicians. Henri Poincaré, whose classic article on "Mathematical Discovery" appears on page 128, was a leading investigator of the subject. The theory of functions of a real variable reached its latest stage with Cantor's theory of sets, which after bitter opposition has gradually been accepted as the foundation of the theory.

The theory of probability, which originated as a guide for gamblers, has had application in communication and statistics. And the logical basis of all mathematics, which began with Boole's "An Investigation of the Laws of Thought" and which Peano elaborated on, had its modern development, as we mention on page 99, in Principia Mathematica by Whitehead and Russell, whose thesis is that all mathematics is an exercise in symbolic logic, and who have thus attempted to place a solid foundation under all mathematical theory.

Various algebraic systems have also been under intensive study in modern times. It has become clear that the manipulation of algebraic symbols rather than their meaning is primarily important (see both Whitehead's "Introduction to Mathematics" and Dantzig's "Symbols"). Groups and rings are algebraic concepts which have received much attention. Of the latter an important class is formed by the hypercomplex numbers, an example of which are the quaternions, discovered by the Irish genius William Rowan Hamilton. Out of a multitude of choices, we have selected the biography of him which follows, partly because of the intrinsic interest of his life, partly because of the importance of his work, and partly because it is written by a famous contemporary mathematician. Sir Edmund Whittaker, the author, studied mathematics at Cambridge under Cayley and Stokes. At Trinity College he was an associate of such mathematicians as Whitehead and Russell, and of such scientists as Thomson and Rutherford. In 1906 he was appointed Professor of Astronomy at Dublin—the chair which had previously been occupied by Hamilton himself—although he spent most of his time in Edinburgh. Sir Edmund died at 83 in 1956.

WILLIAM ROWAN HAMILTON

SIR EDMUND WHITTAKER

AFTER ISAAC NEWTON, the greatest mathematician of the English-speaking peoples is William Rowan Hamilton, who was born in 1805 and died in 1865. His fame has had some curious vicissitudes. During his lifetime he was celebrated but not understood; after his death his reputation declined and he came to be counted in the second rank; in the twentieth century he has become the subject of an extraordinary revival of interest and appreciation.

About his ancestry there is not much to be said. His father was a Dublin solicitor who defended the outlawed Irish patriot Archibald Hamilton Rowan and obtained a reversal of his sentence. From Rowan, who acted as sponsor at the baptism of the infant William, the boy received his second Christian name. The child was not brought up by his own parents. When he was about a year old, they decided to entrust his education to Mr. Hamilton's brother James, a clergyman settled at Trim, a small town 30 miles north of Dublin. Young William lived in Trim, with occasional visits to Dublin, until he was of age to enter the University.

Whether the credit must be given to his uncle's methods of education or to his own natural gifts, it is recorded that by the age of three William could read English easily; at five he was able to read and translate Latin, Greek and Hebrew; at eight he had added Italian and French; before he was 10 he was studying Arabic and Sanskrit. At the age of 14 he wrote a letter in Persian to the Persian ambassador, then on a visit to Dublin.

The boy loved the classics and the poets, but at the age of 15 his interests, and the course of his life, were completely changed when he met one Zerah Colburn, an American youngster who gave an exhibition in Dublin of his powers as a lightning calculator. "For a long time afterwards," wrote Hamilton later, "I liked to perform long operations in arithmetic in my mind; extracting the square and cube root, and everything that related to the properties of numbers." William resolved upon a life of mathematics. "Nothing," he declared: "so exalts the mind, or so raises a man above his fellow-creatures, as the researches of Science. Who would not rather have the fame of Archimedes than that of his conqueror Marcellus? . . . Mighty minds in all ages have combined to rear the vast and beautiful temple of Science, and inscribed their names upon it in imperishable characters; but the edifice is not completed; it is not yet too late to add another pillar or another ornament. I have yet scarcely arrived at its foot, but I may aspire one day to reach its summit."

In his diary there presently appeared such entries as "read Newton's *Life*" and "began Newton's *Principia*." At the age of 16 he made the acquaintance of Laplace's *Mécanique céleste*. (An entry in his journal around this time recounted: "We have been getting up before five for several mornings—that is, my uncle and I; he pulls a string which goes through the wall and is fastened to my shirt at night.") In 1823, preceded by rumors of his intellectual prowess, "Hamilton the prodigy" entered Trinity College at Dublin. There his progress was brilliant, not only on the examinations but also in original research. When he was only 21 years old he submitted to the Royal Irish Academy a paper entitled "A Theory of Systems of Rays" which in effect made a new science of mathematical optics.

In his paper Hamilton's aim was to remodel the geometry of light by establishing one uniform method for the solution of all problems in that science. He started from the already established principles that a ray of light travels by the path that takes the least time (according to the wave theory) or the least "action" (according to the corpuscular theory) in going from one point to another; this is true whether the path is straight or bent by refraction. Hamilton's contribution was to consider the action (or time) as a function of the positions of the points between which the light passes, and to show that this quantity varied when the coordinates of these points varied, according to a law which he called the law of varying action. He demonstrated that all researches on any system of optical rays can be reduced to the study of this single function. Hamilton's discovery of this "characteristic function," as he called it, was an extraordinary achievement of scientific genius. He had originally projected it when he was 16 and he brought it to a form approaching completeness in his twenty-first year.

The communication of the paper was soon followed by a great change in Hamilton's circumstances. The chair of Professor of Astronomy at Trinity College, which paid an annual salary of 250 pounds and conferred on its occupant the title of Royal Astronomer of Ireland, was vacated in 1826 when its holder, the Reverend John Brinkley, was appointed to the Bishopric of Cloyne, once held by the great philosopher George Berkeley. Hamilton was elected as Brinkley's successor a few months later. The election of an undergraduate to a professorial chair was an astonishing event, and it led to some curious consequences. For instance, the Royal Astronomer was by virtue of his office an examiner for the Bishop Law Prize, a mathematical distinction open to candidates of junior bachelor standing, and thus came to pass the anomalous proceeding of an undergraduate examining graduates in the highest branches of mathematics.

While everyone acknowledged the unprecedented honor of Hamilton's appointment to the chair, opinion was sharply divided as to whether he was wise to have accepted it.

In another year or two he would undoubtedly have been elected a fellow of Trinity College, with better financial and other prospects. What determined his choice was the consideration that the royal astronomership was practically a research appointment, involving very little in the way of fixed duties, while a fellow was required to become a clergyman and must soon have developed into a tutor and lecturer, with duties occupying most of his time. To be sure, the research equipment of the astronomical observatory was poor in the extreme, but what really was in the minds both of Hamilton and of the electors was not astronomy but an arrangement by which he could continue the theoretical researches of which the paper on "Systems of Rays" was such a glorious beginning.

Hamilton did have the duty of giving a course of lectures on astronomy. In these it was his custom to discuss the relations of astronomy to physical science in general, to metaphysics and to all related realms of thought. His lectures were so poetic and learned that they attracted crowded audiences of professors and visitors as well as his class of undergraduates; when in 1831 there was some talk of his being transferred to the chair of mathematics, the Board insisted that he remain where he was. As inducement the Board raised his salary to 580 pounds a year and gave him permission to devote his research principally to mathematics.

In 1832 Hamilton announced to the Royal Irish Academy a remarkable discovery in optics which followed up his theory of systems of rays. It had been known for some time that certain biaxial crystals, such as topaz and aragonite, gave rise to two refracted rays, producing a double image. Augustin Fresnel of France had worked out the rules of double refraction. Now Hamilton, investigating by his general method the law of Fresnel, was led to conclude that in certain cases a single ray of incident light in a biaxial crystal should give rise to not merely two but an infinite number of refracted rays, forming a cone, and that in certain other cases a single ray within such a crystal would emerge as a different cone. He therefore proposed from theory two new laws of light, which he called internal and external conical refraction. They were soon verified experimentally by his friend Humphrey Lloyd, a Dublin physicist.

In 1834 Hamilton, then 29, wrote to his uncle: "It is my hope and purpose to remodel the whole of dynamics, in the most extensive sense of the word, by the idea of my characteristic function." He proceeded to apply this principle to the motion of systems of bodies, and in the following year he expressed the equations of motion in a form which showed the duality between the components of momentum of a dynamical system and the coordinates of its position. Only a century later, with the development of quantum theory, did physicists and mathematicians fully realize the importance of this duality.

In 1835 Hamilton received the honor of knighthood, and two years later he was elected president of the Royal Irish Academy. But his private life was less happy. Upon becoming a professor he had set up house with three of his sisters at the Dunsink Observatory on a hill five miles from Dublin. At the age of 26 he fell in love with Helen Maria Bayly, the daughter of a former rector in County Tipperary. She at first refused to entertain his proposal of marriage but ultimately accepted him, and the wedding took place on April 9, 1833. He had remarked in a letter to a friend on her "extreme timidity and delicacy"; these qualities were only too fully confirmed after the marriage. Lady Hamilton bore two sons and a daughter in six years, but she found herself unequal to the work of home administration and left Dunsink for two years to live with a married sister in England. She returned in 1842, but things became no better. Hamilton henceforth had no regular times for his meals, and he began to use alcoholic stimulants to a dangerous extent.

When I held Hamilton's chair, to which I had the honor of being appointed in 1906, many years after his death, I met many people who had known him. The countryside was full of stories about him. One of them concerns his

administration of the 17 acres of farmland around Dunsink
Observatory, of which the Royal Astronomer has control.
Hamilton, who was town-bred, knew nothing of farming,
but in order to supply his household with milk he bought
a cow. After some time, in the ordinary course of nature,
the yield of milk began to fall off. Hamilton went to con-
sult a neighboring farmer. The farmer, knowing with whom
he had to deal, said that the cow, as the solitary occupant
of 17 acres, was suffering from loneliness. Thereupon
Hamilton inquired whether it would be possible to provide
her with companions, and the farmer graciously agreed,
in recognition of a payment by Hamilton, to allow his
cattle to graze on the rich pastures of Dunsink.

In spite of the unfavorable conditions of his life, Hamilton's scientific work went on. In 1843 he made a great
discovery—the calculus of quaternions.

He was led to this discovery by long thought on the
problem of finding a general rule for computing the fourth
proportional to three straight-line segments when the directions of those lines were taken into account. A line
segment with a specified direction is called a vector. It
was well known that a vector in a plane could be represented by a complex number; that is, a number formed
of both real and imaginary numbers, or $x+y\sqrt{-1}$. (The
square root of -1, an imaginary number, is usually written i, so that the expression becomes $x+yi$.) If we represent
real numbers by distances on the x axis of a graph, then
multiplication of any number by -1, changing it to the
negative number, may be thought of as rotating the line
segment 180 degrees, while multiplication by i, the square
root of -1, may be thought of as a 90-degree rotation
(see Figure I-1). Thus imaginary numbers are represented
on the y axis, and i may be considered a unit on that axis,
or a "unit vector." Any vector in a plane may then be
specified by a complex number giving its x and y components. Such a pair of numbers, known as a doublet,
obeys the same algebraic laws as a single number: doublets
can be added, subtracted, multiplied and divided accord-

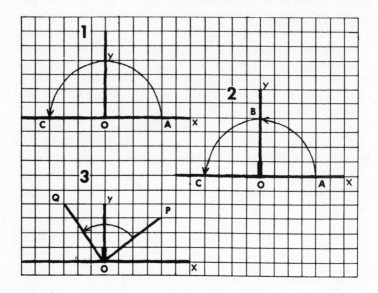

I-1.   Complex number, made up of a real number and an
imaginary one, the square root of −1, is used to describe the
length and direction of a line segment. When complex numbers are added, subtracted or multiplied, the process is equivalent to a geometrical operation, e.g., rotation. In the diagram
numbered 1 the line segment OA, representing the number
+4, is multiplied by −1, which changes it to the line segment
OC, or −4. Thus multiplication by −1 is equivalent to
rotation through 180 degrees. In diagram 2 multiplication by
−1 is done in two steps, i.e., multiplication by $\sqrt{-1}$ and by
$\sqrt{-1}$ again. (The square root of −1 is usually written i)
Consequently multiplication by i can be considered rotation
through 90 degrees. This leads to the idea of measuring
imaginary distances on the y axis, as is indicated by making
i the "unit vector" on that axis. Multiplication by i has the
effect of a 90-degree rotation even if the starting point is not
the x axis. The line segment from point O (x = o, y = o)
to point P (x = 4, y = 3) is represented in complex-number
notation as 4 + 3i. Multiplying this number by i gives 4i
+ 3i^2, or 3 − 4i. The latter number represents the line segment OQ (x = −3, y = 4), or a 90-degree rotation of the
line OP.

ing to the usual rules. Thus it is possible to calculate the fourth proportional to three vectors in a common plane: $V_1: V_2 = V_3: x$.

Hamilton conjectured that in three-dimensional space a vector might be represented by a set of three numbers, a triplet, just as a vector in a plane was expressed by a doublet. He sought to find the fourth proportional by multiplying triplets, but encountered difficulties. The younger members of the household at Dunsink shared affectionately in the hopes and disappointments of their illustrious parent as the investigation proceeded. William Edwin (aged nine) and Archibald Henry (eight) used to ask at breakfast: "Well, Papa, can you multiply triplets?" Whereto he was obliged to reply, with a sad shake of the head, "No, I can only add and subtract them."

One day, while walking from Dunsink into Dublin, Hamilton suddenly realized the answer: the geometrical operations of three-dimensional spaces required for their description not triplets but *quadruplets*. To specify the operation needed to convert one vector into another in space, one had to know four numbers: 1] the ratio of the length of one vector to the other, 2] the angle between them, and 3] the node and 4] the inclination of the plane in which they lie.

Hamilton named the set of four numbers a quaternion, and found that he could multiply quaternions as if they were single numbers. But he discovered that the algebra of quaternions differed from ordinary algebra in a crucial respect: it was *noncommutative*. This word calls for some explanation. When we multiply 2 by 3 we obtain the same product as when we multiply 3 by 2. This *commutative* law of multiplication, as it is called, is embodied in the algebraic formula $ab = ba$. It applies to imaginary numbers as well as to real numbers. It does not, however, hold for the calculus of quaternions, because the latter describes geometrical operations such as rotations. Figure I-2 shows why. It represents three mutually perpendicular axes, the y and z axes lying in the plane of the paper and the x axis extending

toward the reader. The characters *i*, *j* and *k* represent unit
vectors along the x, y and z axes respectively. Multiplication by *i* is defined as a 90-degree counterclockwise rotation
in the plane of the paper; multiplication by *j* and *k*, as rotations in planes perpendicular to that plane. Now multiplication of *j* by *i* rotates *j* to *k*; that is, *ij* = *k*. But multiplication
of *i* by *j* rotates *i* to −*k*. Thus *ij* does not equal *ji*.

The surrender of the commutative law was a tremendous
break with tradition. It marked the beginning of a new era.
The news of the discovery spread quickly, and led, in

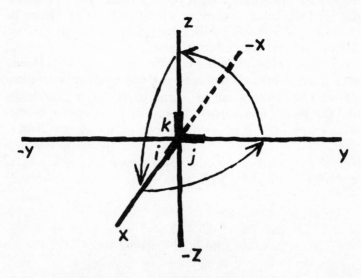

I-2. *Noncommutative algebra is used to represent geometrical
operations in three dimensions. A vector in three dimensions
is represented in a system of coordinates with three mutually
perpendicular axes (x points toward the reader, y and z are
in the plane of the page) in terms of the unit vectors i, j and
k. Multiplication by i is arbitrarily defined as meaning a counterclockwise rotation of 90 degrees in the plane perpendicular
to the i vector, i.e., the plane of y and z. Multiplication by j
and k are similarly defined, as indicated by the arrows. Now it
can be seen that multiplication i × j has the effect of rotating
j into k. On the other hand, the multiplication j × i has the
effect of rotating i into −k. So i × j = k, and j × i = −k. In
other words, the multiplication is noncommutative: i × j does
not equal j × i.*

Dublin at any rate, to a wave of interest among people of rank and fashion like the later boom in general relativity in London, when Lord Haldane invited Einstein to meet the Archbishop of Canterbury at luncheon. Hamilton was buttonholed in the street by members of the Anglo-Irish aristocracy with the question: "What the deuce are the quaternions?" To satisfy them he published the delightful *Letter to a Lady*, in which he explained that the term occurs, for example, in our version of the Bible, where the Apostle Peter is described as having been delivered by Herod to the charge of four quaternions of soldiers. . . . And to take a lighter and more modern instance from the pages of *Guy Mannering*, Scott represents Sir Robert Hazelwood as loading his long sentences with "triads and quaternions."

From this time until his death 22 years later, Hamilton's chief interest was to develop the new calculus. They were mostly sad and lonely years, owing to the frequent illnesses and absences of his wife. He worked all day in the large dining room of the Observatory house, into which from time to time his cook passed a mutton chop. (After his death scores of mutton chop bones on plates were found sandwiched among his papers.)

Hamilton's discovery was quickly followed by other new algebras, such as the theory of matrices, which is likewise noncommutative. Thus he started a glorious school of mathematics, though it was not to come into full flower for another half-century. I remember discussing in 1900 with Alfred North Whitehead whether quaternions and other noncommutative algebras had much of a future as regards applications to physics. Whitehead remarked that while all the physics then known could be treated by ordinary algebra, it was possible that new fields in physics might some day be discovered for which noncommutative algebra would be the only natural representation. In that very year this anticipation was started on the road to fulfillment. Max Planck introduced the quantum h, the beginning of the quantum theory. Now h is a quantum of action, and action was a central conception in Hamilton's system of dynamics. Thus the Hamiltonian ideas on dynamics began to come into prominence. But very slowly. When my book *Analytical Dynamics* was published in 1904, I was criticized severely for devoting a large part of it to such topics as the co-ordinates–momentum duality, action, and other Hamiltonian ideas. The critics called them mere mathematical playthings.

The good work went on, however. The discovery of special relativity brought quaternions to the fore, because Arthur Cayley of Cambridge University had shown in 1854 that quaternions could be applied to the representation of rotations in four-dimensional space. His result yielded a particularly elegant expression for the most general Lorentz transformation. Moreover, the new discoveries again emphasized the importance of action, which preserves its form in different reference systems and is therefore fundamental in relativity physics.

Meanwhile, the workers in quantum theory were coming to realize that Hamilton's dynamical conceptions must form the basis of all rules of quantification. And in 1925 the other side of his work—his noncommutative algebra—was brought into quantum theory by Werner Heisenberg, Max Born, and Pascual Jordan, who showed that the ordinary Hamiltonian equations of dynamics were still valid in quantum theory, provided the symbols representing the coordinates and momenta in classical dynamics were interpreted as operators whose products did not commute.

Time has amply vindicated Hamilton's intuition of the duality between generalized coordinates and generalized momenta. This was strikingly shown in 1927, when Heisenberg discovered the principle of uncertainty, which is usually stated in this way: the more accurately the coordinates of a particle are determined, the less accurately can its momentum be known, and vice versa, the product of the two uncertainties being of the order of Planck's constant h.

Quantum-mechanical workers have generally tended to regard matrices rather than quaternions as the type of non-commutative algebra best suited to their problems, but the original Hamiltonian formulas keep on cropping up. Thus the "spin matrices" of Wolfgang Pauli, on which the quantum-mechanical theory of rotations and angular momenta depends, are simply Hamilton's three quaternion units i, j, k. Arthur Conway has shown that quaternion methods may be used with advantage in the discussion of P. A. M. Dirac's equation for the spinning electron. Hamilton's formulae of 1843 may even yet prove to be the most natural expression of the new physics.

The application of modern mathematics to other studies finds what is probably its most spectacular example in relativity theory, which limits the former inviolability of Newton's laws. Despite the numerous popularizations of this and other physical theories, such as Heisenberg's Principle of Uncertainty, they are basically mathematical concepts, which cannot be truly understood except in mathematical terms. This is so despite the fact that some of the popularizers, such as Eddington and Einstein himself, have been on the highest level.

This limitation should be borne in mind in reading James R. Newman's short discussion which follows. Newman is best known today for his four-volume World of Mathematics, an elaborate and painstaking collection of the best writing in mathematics, which is recommended to the readers who wish to go more deeply into the subject than this volume's space limitations permit. He began his career as a lawyer, teaching at the Yale Law School. He has also acted as an advisor to the government on scientific legislation and is the author, with Edward Kasner, of Mathematics and the Imagination, from which we reprint a selection on page 183.

EINSTEIN'S GREAT IDEA

JAMES R. NEWMAN

> To the eyes of the man of imagination,
> nature is imagination itself.
> —WILLIAM BLAKE

EINSTEIN DIED FOUR YEARS AGO.[1] Fifty years earlier, when he was twenty-six, he put forward an idea which changed the world. His idea revolutionized our conception of the physical universe; its consequences have shaken human society. Since the rise of science in the seventeenth century, only two other men, Newton and Darwin, have produced a comparable upheaval in thought.

Einstein, as everyone knows, did something remarkable, but what exactly did he do? Even among educated men and women, few can answer. We are resigned to the importance of his theory, but we do not comprehend it. It is this circumstance which is largely responsible for the isolation of modern science. This is bad for us and bad for science; therefore more than curiosity is at stake in the desire to understand Einstein.

Relativity is a hard concept, prickly with mathematics. There are many popular accounts of it, a small number of which are good, but it is a mistake to expect they will carry the reader along—like a prince stretched on his palanquin. One must tramp one's own road. Nevertheless, relativity is in some respects simpler than the theory it supplanted. It makes the model of the physical world more susceptible to proof by experiment; it replaces a grandiose scheme of space and time with a more practical scheme. Newton's majestic system was worthy of the gods; Einstein's system is better suited to creatures like ourselves, with limited intelligence and weak eyes.

1. In 1955—*Eds.*

78

But relativity is radically new. It forces us to change deeply rooted habits of thought. It requires that we free ourselves from a provincial perspective. It demands that we relinquish convictions so long held that they are synonymous with common sense, that we abandon a picture of the world which seems as natural and as obvious as that the stars are overhead. It may be that in time Einstein's ideas will seem easy; but our generation has the severe task of being the first to lay the old aside and try the new. Anyone who seeks to understand the world of the twentieth century must make this effort.

In 1905 while working as an examiner in the Swiss Patent Office, Einstein published in the *Annalen der Physik*, a thirty-page paper with the title "On the Electrodynamics of Moving Bodies." The paper embodied a vision. Poets and prophets are not alone in their visions; a young scientist— it happens mostly to the young—may in a flash glimpse a distant peak which no one else has seen. He may never see it again, but the landscape is forever changed. The single flash suffices; he will spend his life describing what he saw, interpreting and elaborating his vision, giving new directions to other explorers.

At the heart of the theory of relativity are questions connected with the velocity of light. The young Einstein began to brood about these while still a high-school student. Suppose, he asked himself, a person could run as fast as a beam of light, how would things look to him? Imagine that he could ride astride the beam, holding a mirror just in front of him. Then, like a fictional vampire, he would cause no image; for since the light and the mirror are traveling in the same direction at the same velocity, and the mirror is a little ahead, the light can never catch up to the mirror and there can be no reflection.

But this applies only to *his* mirror. Imagine a stationary observer, also equipped with a mirror, who watches the rider flashing by. Obviously the observer's mirror will catch the rider's image. In other words, the optical phenomena surrounding this event are purely relative. They exist for the observer; they do not exist for the rider. This was a troublesome paradox, which flatly contradicted the accepted views of optical phenomena. We shall have to see why.

The speed of light had long engaged the attention of physicists and astronomers. In the seventeenth century the Danish astronomer Römer discovered that light needed time for its propagation. Thereafter, increasingly accurate measurements of its velocity were made and by the end of the nineteenth century the established opinion was that light always travels in space at a certain constant rate, about 186,000 miles a second.

But now a new problem arose. In the mechanics of Galileo and Newton, rest and uniform motion (*i.e.*, constant velocity) are regarded as indistinguishable. Of two bodies, A and B, it can only be said that one is in motion *relative* to the other. The train glides by the platform; or the platform glides by the train. The earth approaches the fixed stars; or they approach it. There is no way of deciding which of these alternatives is true. And in the science of mechanics it makes no difference.

One of the questions, therefore, was whether, in respect to motion, light itself was like a physical body; that is, whether its motion was relativistic in the Newtonian sense, or absolute.

The wave theory of light appeared to answer this question. A wave is a progressive motion in some kind of medium; a sound wave, for example, is a movement of air particles. Light waves, it was supposed, move in an all-pervasive medium called the ether. The ether was assumed to be a subtle jelly with marvelous properties. It was colorless, odorless, without detectable features of any kind. It could penetrate all matter. It quivered in transmitting light. Also, the body of the ether as a whole was held to be stationary. To the physicist this was its most important property, for being absolutely at rest the ether offered a unique frame of reference for determining the velocity of light. Thus while it was hopeless to attempt to determine the absolute motion of a physical body because one could find no absolutely stationary frame of reference against which to measure it, the attempt was not hopeless for light; the ether, it was thought, met the need.

The ether, however, did not meet the need. Its marvelous properties made it a terror for experimentalists. How could motion be measured against an ectoplasm, a substance with no more substantiality than an idea? Finally, in 1887, two American physicists, A. A. Michelson and E. W. Morley, rigged up a beautifully precise instrument, called an interferometer, with which they hoped to discover some evidence of the relationship between light and the hypothetical ether. If the earth moves through the ether, a beam of light traveling in the direction of the earth's motion should move faster through the ether than a beam traveling in the opposite direction. Moreover, just as one can swim across a river and back more quickly than one can swim the same distance up and down stream, it might be expected that a beam of light taking analogous paths through the ether would complete the to-and-fro leg of the journey more quickly than the up-and-down leg.

This reasoning was the basis of the Michelson-Morley experiment. They carried out a number of trials in which they compared the velocity of a beam of light moving through the ether in the direction of the earth's motion, and another beam traveling at right angles to this motion. There was every reason to believe that these velocities would be different. Yet no difference was observed. The light beam seemed to move at the same velocity in either direction. The possibility that the earth dragged the ether with it having been ruled out, the inquiry had come to a dead end. Perhaps there was no difference; perhaps there was no ether. The Michelson-Morley findings were a major paradox.

Various ideas were advanced to resolve it. The most imaginative of these, and also the most fantastic, was put forward by the Irish physicist, G. F. Fitzgerald. He suggested that since matter is electrical in essence and held together by electrical forces, it may contract in the direction of its motion as it moves through the ether. The contraction would be very small; nevertheless in the direction of motion the unit of length would be shorter. This hypothesis would explain the Michelson-Morley result. The arms of their interferometer might contract as the earth rotated; this would shorten the unit of length and cancel out the added velocity imparted to the light by the rotation of the earth. The velocities of the two beams—in the direction of the earth's motion and at right angles to it—would appear equal. Fitzgerald's idea was elaborated by the famous Dutch physicist, H. A. Lorentz. He put it in mathematical form and connected the contraction caused by motion with the velocity of light. According to his arithmetic, the contraction was just enough to account for the negative results of the Michelson-Morley experiment. There the subject rested until Einstein took it up anew.

He knew of the Michelson-Morley findings. He knew also of other inconsistencies in the contemporary model of the physical world. One was the slight but persistent misbehavior (by classical standards) of the planet Mercury as it moved in its orbit; it was losing time (at a trifling rate, to be sure—forty-three seconds of arc per century), but Newton's theory of its motion was exact and there was no way of accounting for the discrepancy. Another was the bizarre antics of electrons, which, as W. Kaufmann and J. J. Thomson discovered, increased in mass as they went faster. The question was, could these inconsistences be overcome by patching and mending classical theories? Or had the time come for a Copernican renovation?

Making his own way, Einstein turned to another aspect of the velocity problem. Velocity measurements involve time measurements, and time measurements, as he perceived, involve the concept of simultaneity. Is this concept simple and intuitively clear? No one doubted that it was; but Einstein demanded proof.

I enter my study in the morning as the clock on the wall begins to strike. Obviously these events are simultaneous. Assume, however, that on entering the study I hear the first stroke of the town-hall clock, half a mile away. It took time for the sound to reach me; therefore while the sound wave fell on my ears at the moment I entered the study, the event that produced the wave was not simultaneous with my entry.

Consider another kind of signal. I see the light from a distant star. An astronomer tells me that the image I see is not of the star as it is today, but of the star as it was the year Brutus killed Caesar. What does simultaneity mean in this case? Is my *here-now* simultaneous with the star's *there-then*? Can I speak meaningfully of the star as it was the day Joan of Arc was burned, even though ten generations will have to pass before the light emitted by the star on that day reaches the earth? How can I be sure it will ever get here? In short, is the concept of simultaneity for different places exactly equivalent to the concept for one and the same place?

Einstein soon convinced himself that the answer is no. Simultaneity, as he realized, depends on signals; the speed of light (or other signal) must therefore enter into the meaning of the concept. Not only does the separation of events in *space* becloud the issue of simultaneity in *time*, but relative motion may do so. A pair of events which one observer pronounces simultaneous may appear to another observer, in motion with respect to the first, to have happened at different times. In his own popular account of relativity (see Figure I-3), Einstein gave a convincing and easy example, which showed that *any* measurement of time is a measurement with respect to a given observer. A measurement valid for one observer may not be valid for another. Indeed, the measurement is certain not to be valid if one attempts to extend it from the system where the measurement was made to a system in motion relative to the first.

Einstein was now aware of these facts. Measuring the speed of light requires a time measurement. This involves a judgment of simultaneity. Simultaneity is not an absolute fact, the same for all observers. The individual observer's judgment depends on relative motion.

But the sequence does not end here. A further inference suggests itself, namely, that simultaneity may also be involved in measuring distances. A passenger on a moving train who wants to measure the length of his car has no difficulty. With a yardstick he can do the job as easily as if

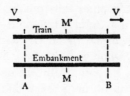

I-3.   *Einstein's Own Example of the Relativity of Time*
The diagram shows a long railroad train traveling along the
rails with Velocity V, in the direction toward the right of the
page. The bottom line denotes the embankment running
parallel to the rails. The letters A and B mark two places on
the rails, and the letter M marks a point on the embankment
directly midway between A and B. At M stands an observer
equipped with a pair of mirrors which are joined in a V and
inclined at 90°. By means of the device he can observe both
places, A and B, at the same time. We imagine two events at
A and B, say two flashes of lightning, which the observer
perceives in his mirror device at the same time. These he
pronounces to be simultaneous, by which he means that the
rays of light emitted at A and B by the lightning bolts meet
at the midpoint M of the length A → B along the embankment. Now consider the moving train, and imagine a passenger
seated in it. As the train proceeds along the rails, the passenger
will arrive at a point M', which is directly opposite M, and
therefore exactly midway between the length A → B along
the rails. Assume further that the passenger arrives at M' just
when the flashes of lightning occur. We have seen that the
observer at M correctly pronounces the lightning bolts as
simultaneous; the question is, will the train passenger at M'
make the same pronouncement? It is easily shown that he will
not. Obviously if the point M' were stationary with respect
to M, the passenger would have the same impression of simultaneity of the lightning flashes as the observer on the embankment. But M' is not stationary; it is moving toward the right
with the velocity V of the train. Therefore (considered with
reference to the embankment) the passenger is moving toward
the beam of light coming from B, and away from the beam
coming from A. It seems clear then that he will see the beam
emitted by the flash at B sooner than the beam emitted by the
flash at A. Accordingly he will pronounce the flash at B as
earlier in time than the flash at A.

    Which of the two pronouncements is correct, the observer's
or the passenger's? The answer is that each is right in its
own system. The observer is right with respect to the embankment, the passenger with respect to the train. The observer may say that he alone is right because he is at rest
while the passenger is moving and his impressions are there-

he were measuring his room at home. Not so for a stationary observer watching the train go by. The car is moving and he cannot measure it simply by laying a yardstick end on end. He must use light signals, which will tell him when the ends of the car coincide with certain arbitrary points. Therefore, problems of time arise. Suppose the thing to be measured is an electron, which is in continual motion at high speed. Light signals will enter the experiment, judgments of simultaneity will have to be made, and once again it is obvious that observers of the electron who are in motion relative to each other will get different results. The whole comfortable picture of reality begins to disintegrate: neither space nor time is what it seems.

The clarification of the concept of simultaneity thrust upon Einstein the task of challenging two assumptions, assumptions hedged with the divinity of Isaac Newton. "Absolute, true, and mathematical time, of itself and from its own nature, flows equably without relation to anything external. . . ." This was Newton's sonorous definition in his great book, *Principia Mathematica*. To this definition he added the equally majestic, "Absolute space, in its own nature, without relation to anything external, remains always similar and immovable." These assumptions, as Einstein saw, were magnificent but untenable. They were at the bottom of the paradoxes of contemporary physics. They had to be discarded. Absolute time and absolute space were concepts which belonged to an outworn metaphysic. They went

---

fore distorted. To this the passenger can reply that motion does not distort the signals, and that, in any case, there is no more reason to believe that he was moving and the observer at rest than that the passenger was at rest and the observer moving.

There is nothing to choose between these views, and they can be logically reconciled only by accepting the principle that simultaneity of events is meaningful only with respect to a particular reference system; moreover, that every such system has its own particular time, and unless, as Einstein says, we are told the reference system to which the statement of time refers, a bare statement of the time of an event is meaningless.

beyond observation and experiment; indeed, they were refuted by the nasty facts. Physicists had to live with these facts.

To live with them meant nothing less than to accept the Michelson-Morley paradox, to incorporate it into physics rather than try to explain it away. From the point of view of common sense the results were extraordinary, yet they had been verified. It was not the first time that science had had to overrule common sense. The evidence showed that the speed of light measured by *any* observer, whether at rest or in motion relative to the light source, is the same. Einstein embodied this fact in a principle from which a satisfactory theory of the interaction between the motion of bodies and the propagation of light could be derived. This principle, or first postulate, of his Special Theory of Relativity states that *the velocity of light in space is a constant of nature, unaffected by the motion of the observer or of the source of the light.*

The hypothesis of the ether thus became unnecessary. One did not have to try to measure the velocity of light against an imaginary frame of reference, for the plain reason that whenever light is measured against *any* frame of reference its velocity is the same. Why then conjure up ethereal jellies? The ether simply lost its reason for being.

A second postulate was needed. Newtonian relativity applied to the motion of material bodies; but light waves, as I mentioned earlier, were thought not to be governed by this principle. Einstein pierced the dilemma in a stroke. He simply extended Newtonian relativity to include optical phenomena. The second postulate says: *In any experiment involving mechanical or optical phenomena it makes no difference whether the laboratory where the experiment is being performed is at rest or in uniform motion; the results of the experiments will be the same.* More generally, one cannot by any method distinguish between rest and uniform motion, except in relation to each other.

Is that all there is to the special theory of relativity? The postulates are deceptively simple. Moreover, to the sharp-eyed reader they may appear to contradict each other. The contradictions, however, are illusory, and the consequences are revolutionary.

Consider the first point. From the postulates one may infer that on the one hand light has the velocity c, and, on the other hand, even when according to our traditional way of calculating it should have the velocity $c + q$ (where q is the velocity of the source), its velocity is still c. Concretely, light from a source in motion with respect to a given frame of reference has the same velocity as light from a source at rest with respect to the same frame. (As one physicist suggested, this is as if we were to say that a man walking up a moving stairway does not get to the top any sooner than a man standing still on the moving stairway.) This seems absurd. But the reason it seems absurd is that we take it for granted that the velocity of the moving source must be added to the normal velocity of light to give the correct velocity of the beam emitted by the source. Suppose we abandon this assumption. We have already seen, after all, that motion has a queer effect on space and time measurements. It follows that the established notions of velocity must be reconsidered. The postulates were not inherently contradictory; the trouble lay with the classical laws of physics. They had to be changed. Einstein did not hesitate. To preserve his postulates he consigned the old system to the flames. In them were consumed the most cherished notions of space, time and matter.

One of the clichés about Einstein's theory is that it shows that everything is relative. The statement that everything is relative is as meaningful as the statement that everything is bigger. As Bertrand Russell pointed out, if everything were relative there would be nothing for it to be relative to. The name relativity is misleading. Einstein was in fact concerned with finding something that is *not* relative, something that mathematicians call an invariant. With this as a fixed point, it might be possible to formulate physical laws which would incorporate the "objective residue" of an observer's experience; that is, that part of the space and time characteristics of a physical event which, though perceived by him, are independent of the observer and might therefore be expected to appear the same to all observers. The constancy principle of the velocity of light provided Einstein with the invariant he needed. It could be maintained, however, only at the expense of the traditional notion of time. And even this offering was not enough. Space and time are intertwined. They are part of the same reality. Tinkering with the measure of time unavoidably affects the measure of space.

Einstein, you will notice, arrived at the same conclusions as Fitzgerald and Lorentz without adopting their electrical hypotheses. It was a consequence of his postulates that clocks and yardsticks yield different measurements in relative motion than at rest. Is this due to an actual physical change in the instruments? The question may be regarded as irrelevant. The physicist is concerned only with the difference in measurements. If clock springs and yardsticks contract, why is it not possible to detect the change? Because any scales used to measure it would suffer the same contraction. What is at issue is nothing less than the foundations of rational belief.

Earlier I mentioned Kaufmann's and Thomson's discovery that a moving electron increases in mass as it goes faster. Relativity explains this astonishing fact. The first postulate sets an upper limit to the velocity of light, and permits of the deduction that no material body can exceed this speed limit. In Newton's system there were no such limits; moreover, the mass of a body—which he defined as its "quantity of matter"—was held to be the same whether the body was at rest or in motion. But just as his laws of motion have been shown not to be universally true, his concept of the constancy of mass turns out to be flawed. According to Einstein's Special Theory, the resistance of a body to changes in velocity increases with velocity. Thus, for example, more force is required to increase a body's velocity from 50,000 to 50,001 miles per hour than from 100 to 101 miles per hour. The scientific name for this resistance is *inertia*, and the measure of inertia is mass. (This jibes with the intuitive notion that the amount of force needed to accelerate a body depends on its "quantity of matter.") The ideas fall neatly into place: with increased speed, inertia increases; increased inertia evinces itself as increased mass. The increase in mass is, to be sure, very small at ordinary speeds, and therefore undetectable, which explains why Newton and his successors, though a brilliant company, did not discover it. This circumstance also explains why Newton's laws are perfectly valid for all ordinary instances of matter in motion: even a rocket moving at 10,000 miles an hour is a tortoise compared to a beam of light at 186,000 miles a second. But the increase in mass becomes a major factor where high-speed nuclear particles are concerned; for example, the electrons in a hospital X-ray tube are speeded up to a point where their normal mass is doubled, and in an ordinary TV-picture tube the electrons have 5 per cent extra mass due to their energy of motion. And at the speed of light the push of even an unlimited accelerating force against a body is completely frustrated, because the mass of the body, in effect, becomes infinite.

It is only a step now to Einstein's fateful mass-energy equation.

The quantity of additional mass, multiplied by an enormous number—namely, the square of the speed of light—is equivalent to the energy which was turned into mass. But is this equivalence of mass and energy a special circumstance attendant upon motion? What about a body at rest? Does its mass also represent energy? Einstein boldly concluded that it does. "The mass of a body is a measure of its energy content," he wrote in 1905, and gave his now-famous formula, $E = mc^2$, where E is energy content, m is mass (which varies according to speed) and c is the velocity of light.

"It is not impossible," Einstein said in this same paper, "that with bodies whose energy content is variable to a high degree (*e.g.*, with radium salts) the theory may be successfully put to the test." In the 1930's many physicists were making this test, measuring atomic masses and the energy of products of many nuclear reactions. All the results verified his idea. A distinguished physicist, Dr. E. U. Condon, tells a charming story of Einstein's reaction to this triumph: One of my most vivid memories is of a seminar at Princeton (1934) when a graduate student was reporting on researches of this kind and Einstein was in the audience. Einstein had been so preoccupied with other studies that he had not realized that such confirmation of his early theories had become an everyday affair in the physical laboratory. He grinned like a small boy and kept saying over and over, *"Is das wirklich so?"* Is it really true? —as more and more specific evidence of his $E = mc^2$ relation was being presented.

For ten years after he formulated the Special Theory, Einstein grappled with the task of generalizing relativity to include accelerated motion. This article cannot carry the weight of the details, but I shall describe the matter briefly.

While it is impossible to distinguish between rest and uniform motion by observations made within a system, it seems quite possible, under the same circumstances, to determine *changes* in velocity or direction, *i.e.*, acceleration. In a train moving smoothly in a straight line, at constant velocity, one feels no motion. But if the train speeds up, slows down or takes a curve, the change is felt immediately. One has to make an effort to keep from falling, to prevent the soup from sloshing out of the plate, and so on. These effects are ascribed to what are called *inertial forces*, producing acceleration—the name is intended to convey the fact that the forces arise from the inertia of a mass, *i.e.*, its resistance to changes in its state. It would seem then that any one of several simple experiments should furnish evidence of such acceleration, and distinguish it from uniform motion or rest. Moreover, it should even be possible to determine the effect of acceleration on a beam of light. For example, if a beam were set parallel to the floor of a laboratory at rest or in uniform motion, and the laboratory were accelerated upward or downward, the light would no longer be parallel to the floor, and by measuring the deflection one could compute the acceleration.

When Einstein turned these points over in his mind, he perceived a loose end in the reasoning, which others had not noticed. How is it possible in either a mechanical or an optical experiment to distinguish between the effects of gravity, and of acceleration produced by inertial forces? Take the light-beam experiment. At one point the beam is parallel to the floor of the laboratory; then suddenly it is deflected. The observer ascribes the deflection to acceleration caused by inertial forces, but how can he be sure? He must make his determination entirely on the basis of what he sees *within* the laboratory, and he is therefore unable to tell whether inertial forces are at work—as in the moving train—or whether the observed effects are produced by a large (though unseen) gravitating mass.

Here then, Einstein realized, was the clue to the problem of generalizing relativity. As rest and uniform motion are indistinguishable, so are acceleration and the effects of gravitation. Neither mechanical nor optical experiments conducted within a laboratory can decide whether the system is accelerated or in uniform motion and subjected to a gravitational field. (The poor wretch in tomorrow's space ship, suddenly thrown to the floor, will be unable to tell whether his vehicle is starting its rocket motors or a huge gravitational mass has suddenly appeared to bedevil it.) Einstein formulated his conclusion in 1911 in his "principle of equivalence of gravitational forces and inertial forces."

His ideas invariably had startling consequences. From the principle of equivalence he deduced, among others, that gravity must affect the path of a ray of light. This follows from the fact that acceleration would affect the ray, and gravity is indistinguishable from acceleration. Einstein predicted that this gravity effect would be noticeable in the deflection of the light from the fixed stars whose rays pass close to the huge mass of the sun. He realized, of course, that it would not be easy to observe the bending because under ordinary conditions the sun's brilliant light washes out the light of the stars. But during a total eclipse the stars near the sun would be visible, and circumstances would be favorable to checking his prediction. "It would be extremely desirable," Einstein wrote in his paper enunciating the equivalence principle, "if astronomers would look into the problem presented here, even though the consideration developed above may appear insufficiently founded or even bizarre." Eight years later, in 1919, a British eclipse expedition headed by the famous astronomer Arthur Eddington, confirmed Einstein's astounding prediction.

In 1916 Einstein announced his General Theory of Relativity, a higher synthesis incorporating both the Special Theory and the principle of equivalence. Two profound ideas are developed in the General Theory: the union of time and space into a four-dimensional continuum (a consequence of the Special Theory), and the curvature of space.

It was to one of his former professors at Zurich, the Russian-born mathematician, Hermann Minkowski, that Einstein owed the idea of the union of space and time. "From henceforth," Minkowski had said in 1908, "space in itself and time in itself sink to mere shadows, and only a kind of union of the two preserves an independent existence." To the three familiar dimensions of space, a fourth, of time, had to be added, and thus a single new medium, space-time, replaced the orthodox frame of absolute space and absolute time. An event within this medium—one may, for example, think of a moving object as an "event"—is identified not only by three spatial coordinates denoting *where* it is, but by a time coordinate denoting *when* the event is there. *Where* and *when* are, as we have seen, judgments made by an observer, depending on certain interchanges of light signals. It is for this reason that the time coordinate includes as one of its elements the number for the velocity of light.

With absolute space and time discarded, the old picture of the universe proceeding moment by moment from the past through the present into the future must also be discarded. In the new world of Minkowski and Einstein, there is neither absolute past nor absolute future; nor is there an absolute present dividing past from future and "stretching everywhere at the same moment through space." The motion of an object is represented by a line in space-time called a "world-line." The event makes its own history. The signals it emits take time to reach the observer; since he can record only what he sees, an event present for one observer may be past for another, future for a third. In Eddington's words, the absolute "here-now" of former beliefs has become a merely relative "seen-now."

But this must not be taken to mean that every observer can portray only his own world, and that in place of Newtonian order we have Einsteinian anarchy. Just as it was possible in the older sense to fix precisely the distance between two points in three-dimensional space, so it is possible in the four-dimensional continuum of space-time to define and measure distance between events. This distance is called an "interval" and has a "true, absolute value," the same for all who measure it. Thus, after all, "we have found something firm in a shifting world."

How is the concept of curved space related to this picture? The concept itself sticks in the craw. A vase, a pretzel, a line can be curved. But how can empty space be curved? Once again we must think not in terms of metaphysical abstractions, but of testable concepts.

Light rays in empty space move in straight lines. Yet in some circumstances (e.g., where the ray is close to the sun) the path of motion is seen to be curved. A choice of explanations offers itself. We may, for example, say that a gravitational mass in the neighborhood of the ray has bent it; or we may say that this gravitational mass has curved the space through which the ray is traveling. There is no logical reason to prefer one explanation to the other. Gravitational fields are no less an imaginary concept than space-time. The only concrete evidence comes from measuring the path of the light itself—not the field or space-time. It turns out to be more fruitful to explain the curved path of the light ray as an effect of curved space-time, rather than as an effect of the direct action of gravity on light.

Let me suggest an analogy. A thin sheet of rubber is stretched over a large drum-kettle. I take a very light marble and permit it to roll over the sheet. I observe that the path of its motion is a straight line. I now take several lead weights and place them at different points on the rubber sheet. Their weight dimples it, forming small slopes and hollows. Suppose I release the marble on this surface. The path of motion will no longer be straight, but will curve toward the slopes and eventually fall into one of the hollows. Now think of space-time as corresponding to the sheet of rubber, and large gravitational masses to the lead weights; think also of any "event"—a moving particle, a beam of light, a planet—as the counterpart of the marble rolling on the membrane. Where there are no masses, space-time is "flat" and paths of motion are straight lines. But in the neighborhood of large masses space-time is distorted into "slopes" and "hollows," which affect the path of any object entering upon them.

This is what used to be called the attraction of gravitation. But gravitation in Einstein's theory is merely an aspect of space-time. The starlight bent toward the sun "dips" into the "slope" around it, but has enough energy not to be trapped in the "hollow"; the earth circling the sun is riding on the "rim" of its "hollow" like a cyclist racing round a velodrome; a planet which gets too deep into the "hollow" may fall to the bottom. (This is one of the hypotheses astronomers make about collisions which may have formed new planets in our universe.) There are slopes and hollows wherever there is matter; and since astronomical evidence seems to favor the hypothesis that matter is on average uniformly distributed throughout the universe, and finite— though not necessarily constant—Einstein suggested the possibility that the whole of space-time is gently curved, finite, but unbounded. It is not inconsistent with this hypothesis that the universe is expanding, in which case the density of matter would decrease. A finite but unbounded universe is roughly analogous—though it is of higher dimension—to the two-dimensional curved surface of the earth. The area is finite without boundaries, and if one travels in a "straight line" in a given direction one must, after a time, return to the original point of departure.

Einstein's achievement is one of the glories of man. Two points about his work are worth making. The first is that his model of the world was not a machine with man outside it as observer and interpreter. The observer is part of the reality he observes; therefore by observation he shapes it.

The second point is that his theory did much more than answer questions. As a living theory it forced new questions upon us. Einstein challenged unchallengeable writs; he would have been the last to claim that his own writs were beyond challenge. He broadened the human mind.

II. MATHEMATICS and the Mathematician

II. Mathematics and the Mathematician

In the year 1900, two Trinity College mathematicians, Bertrand Russell and Alfred North Whitehead, attended the International Congress of Philosophy in Paris to hear Giuseppe Peano of the University of Turin lecture on symbolic logic. Concerning this experience, Russell was to write, "The great master of formal reasoning, among the men of our own day . . . has reduced the greater part of mathematics . . . to strict symbolic form, in which there are no words at all. In the ordinary mathematics books, there are no doubt fewer words than most readers would wish. Still, little phrases occur, such as therefore, let us assume, consider, or hence it follows. All these, however, are a concession, and are swept away by Professor Peano." The lectures were profoundly stimulating to the pair. During a decade of intensive study, they elaborated on Peano's ideas. The results were presented in one of the classics of modern mathematics, Principia Mathematica. Of this long and difficult work, Tobias Dantzig relates the quip that probably only two persons, presumably the authors, have read it in its entirety.

Alfred North Whitehead was born at Ramsgate, England, in 1861 and died at Cambridge, Mass., in 1947. He studied at Trinity and became a Fellow of the College, specializing in mathematics. His first book, A Treatise on Universal Algebra, which appeared in 1898, earned him an international reputation, and with other achievements resulted in his election to the Royal Academy. His great contribution to the study of mathematics was his collaboration with Russell. But he was also the author of the charming Introduction to Mathematics, which was first published in 1911, and which has gone through many editions.

Whitehead had distinguished careers, not only as a mathematician, but also as a teacher and philosopher. He gathered around him a group of brilliant friends and students, not only at Cambridge, England, but also at Harvard, where he became

Professor of Philosophy in 1924. They delighted in the richness of his personality and conversation, although they found themselves puzzled by his philosophical ideas. His cosmology remains a matter of controversy to the rare few who profess to understand it, and Whitehead himself was sufficiently simple and modest to have no quarrel with his critics.

An Introduction to Mathematics is recognized as being among the finest of works in its field. The following selection from it carries us to the heart of the matter. It presents the germ of the idea on which he elaborated in his more technical works.

THE NATURE OF MATHEMATICS

ALFRED NORTH WHITEHEAD

The Abstract Nature of Mathematics

THE STUDY OF MATHEMATICS is apt to commence in disappointment. The important applications of the science, the theoretical interest of its ideas, and the logical rigor of its methods, all generate the expectation of a speedy introduction to processes of interest. We are told that by its aid the stars are weighed and the billions of molecules in a drop of water are counted. Yet, like the ghost of Hamlet's father, this great science eludes the efforts of our mental weapons to grasp it—" 'Tis here, 'tis there, 'tis gone"—and what we do see does not suggest the same excuse for illusiveness as sufficed for the ghost, that it is too noble for our gross methods. "A show of violence," if ever excusable, may surely be "offered" to the trivial results which occupy the pages of some elementary mathematical treatises.

The reason for this failure of the science to live up to its reputation is that its fundamental ideas are not explained to the student disentangled from the technical procedure which has been invented to facilitate their exact presentation in particular instances. Accordingly, the unfortunate learner finds himself struggling to acquire a knowledge of a mass of details which are not illuminated by any general conception. Without a doubt, technical facility is a first requisite for valuable mental activity: we shall fail to appreciate the rhythm of Milton, or the passion of Shelley, so long as we find it necessary to spell the words and are not quite certain of the forms of the individual letters. In this sense there is no royal road to learning. But it is equally an error to confine attention to technical processes, excluding consideration of general ideas. Here lies the road to pedantry.

The first acquaintance which most people have with mathematics is through arithmetic. That two and two make four is usually taken as the type of a simple mathematical proposition which everyone will have heard of. Arithmetic, therefore, will be a good subject to consider in order to discover, if possible, the most obvious characteristic of the science. Now, the first noticeable fact about arithmetic is that it applies to everything, to tastes and to sounds, to apples and to angels, to the ideas of the mind and to the bones of the body. The nature of the things is perfectly indifferent, of all things it is true that two and two make four. Thus we write down as the leading characteristic of mathematics that it deals with properties and ideas which are applicable to things just because they are things, and apart from any particular feelings, or emotions, or sensations, in any way connected with them. This is what is meant by calling mathematics an abstract science.

The result which we have reached deserves attention. It is natural to think that an abstract science cannot be of much importance in the affairs of human life, because it has omitted from its consideration everything of real interest. It will be remembered that Swift, in his description of Gulliver's voyage to Laputa, is of two minds on this point. He describes the mathematicians of that country as silly and useless dreamers, whose attention has to be awakened by flappers. Also, the mathematical tailor measures his height by a quadrant, and deduces his other dimensions by a rule and compasses, producing a suit of very ill-fitting clothes. On the other hand, the mathematicians of Laputa, by their marvellous invention of the magnetic island floating in the air, ruled the country and maintained their ascendancy over their subjects. Swift, indeed, lived at a time peculiarly unsuited for gibes at contemporary mathematicians. Newton's *Principia* had just been written, one of the great forces which have transformed the modern world. Swift might just as well have laughed at an earthquake.

But a mere list of the achievements of mathematics is an unsatisfactory way of arriving at an idea of its importance. It is worthwhile to spend a little thought in getting at the root reason why mathematics, because of its very abstractness, must always remain one of the most important topics for thought. Let us try to make clear to ourselves why explanations of the order of events necessarily tend to become mathematical.

Consider how all events are interconnected. When we see the lightning, we listen for the thunder; when we hear the wind, we look for the waves on the sea; in the chill autumn, the leaves fall. Everywhere order reigns, so that when some circumstances have been noted we can foresee that others will also be present. The progress of science consists in observing these interconnections and in showing with a patient ingenuity that the events of this ever-shifting world are but examples of a few general connections or relations called laws. To see what is general in what is particular and what is permanent in what is transitory is the aim of scientific thought. In the eye of science, the fall of an apple, the motion of a planet round a sun, and the clinging of the atmosphere to the earth are all seen as examples of the law of gravity. This possibility of disentangling the most complex evanescent circumstances into various examples of permanent laws is the controlling idea of modern thought.

Now let us think of the sort of laws which we want in order completely to realize this scientific ideal. Our knowledge of the particular facts of the world around us is gained from our sensations. We see, and hear, and taste, and smell, and feel hot and cold, and push, and rub, and ache, and tingle. These are just our own personal sensations: my toothache cannot be your toothache, and my sight cannot be your sight. But we ascribe the origin of these sensations to relations between the things which form the external world. Thus the dentist extracts not the toothache but the tooth. And not only so, we also endeavour to imagine the world as one connected set of things which underlies all the perceptions of all people. There is not one world of things for my sensations and another for yours, but one world in which we both exist. It is the same tooth both for dentist and patient. Also we hear and we touch the same world as we see.

It is easy, therefore, to understand that we want to describe the connections between these external things in some way which does not depend on any particular sensations, nor even on all the sensations of any particular person. The laws satisfied by the course of events in the world of external things are to be described, if possible, in a neutral universal fashion, the same for blind men as for deaf men, and the same for beings with faculties beyond our ken as for normal human beings.

But when we have put aside our immediate sensations, the most serviceable part—from its clearness, definiteness, and universality—of what is left is composed of our general ideas of the abstract formal properties of things; in fact, the abstract mathematical ideas mentioned above. Thus it comes about that, step by step, and not realizing the full meaning of the process, mankind has been led to search for a mathematical description of the properties of the universe, because in this way only can a general idea of the course of events be formed, freed from reference to particular persons or to particular types of sensation. For example, it might be asked at dinner: "What was it which underlay my sensation of sight, yours of touch, and his of taste and smell?" the answer being "an apple." But in its final analysis, science seeks to describe an apple in terms of the positions and motions of molecules, a description which ignores me and you and him, and also ignores sight and touch and taste and smell. Thus mathematical ideas, because they are abstract, supply just what is wanted for a scientific description of the course of events.

This point has usually been misunderstood, from being thought of in too narrow a way. Pythagoras had a glimpse of it when he proclaimed that number was the source of all things. In modern times the belief that the ultimate explanation of all things was to be found in Newtonian mechanics was an adumbration of the truth that all science as it grows towards perfection becomes mathematical in its ideas.

Variables

Mathematics as a science commenced when first someone, probably a Greek, proved propositions about *any* things or about *some* things, without specification of definite particular things. These propositions were first enunciated by the Greeks for geometry; and, accordingly, geometry was the great Greek mathematical science. After the rise of geometry centuries passed away before algebra made a really effective start, despite some faint anticipations by the later Greek mathematicians.

The ideas of *any* and of *some* are introduced into algebra by the use of letters, instead of the definite numbers of arithmetic. Thus, instead of saying that $2+3=3+2$, in algebra we generalize and say that, if x and y stand for *any* two numbers, then $x+y=y+x$. Again, in the place of saying that $3>2$, we generalize and say that if x be *any* number there exists *some* number (or numbers) y such that $y>x$. We may remark in passing that this latter assumption—for when put in its strict ultimate form it is an assumption—is of vital importance, both to philosophy and to mathematics; for by it the notion of infinity is introduced. Perhaps it required the introduction of the arabic numerals, by which the use of letters as standing for definite numbers has been completely discarded in mathematics, in order to suggest to mathematicians the technical convenience of the use of letters for the ideas of *any* number and *some* number. The

Romans would have stated the number of the year in which this is written in the form MDCCCCX, whereas we write it 1910, thus leaving the letters for the other usage. But this is merely a speculation. After the rise of algebra the differential calculus was invented by Newton and Leibnitz, and then a pause in the progress of the philosophy of mathematical thought occurred so far as these notions are concerned; and it was not till within the last few years that it has been realized how fundamental *any* and *some* are to the very nature of mathematics, with the result of opening out still further subjects for mathematical exploration.

Let us now make some simple algebraic statements, with the object of understanding exactly how these fundamental ideas occur.

1]   For *any* number x, $x+2=2+x$;
2]   For *some number* x, $x+2=3$;
3]   For *some* number x, $x+2>3$.

The first point to notice is the possibilities contained in the meaning of *some*, as here used. Since $x+2=2+x$ for any number x, it is true for *some* number x. Thus, as here used, *any* implies *some* and *some* does not exclude *any*. Again, in the second example, there is, in fact, only one number x, such as $x+2=3$, namely only the number 1. Thus the *some* may be that one number only. But in the third example, any number x which is greater than 1 gives $x+2>3$. Hence there are an infinite number of numbers which answer to the *some* number in this case. Thus *some* may be anything between *any* and *one only*, including both these limiting cases.

It is natural to supersede the statements 2] and 3] by the questions:

2']   For what number x is $x+2=3$;
3']   For what numbers x is $x+2>3$.

Considering 2'], $x+2=3$ is an equation, and it is easy to see that its solution is $x=3-2=1$. When we have asked the question implied in the statement of the equation $x+2=3$,

x is called the unknown. The object of the solution of the equation is the determination of the unknown. Equations are of great importance in mathematics, and it seems as though 2'] exemplified a much more thoroughgoing and fundamental idea than the original statement 2]. This, however, is a complete mistake. The idea of the undetermined "variable" as occurring in the use of "some" or "any" is the really important one in mathematics; that of the "unknown" in an equation, which is to be solved as quickly as possible, is only of subordinate use, though of course it is very important. One of the causes of the apparent triviality of much of elementary algebra is the preoccupation of the textbooks with the solution of equations. The same remark applies to the solution of the inequality 3'] as compared to the original statement 3].

But the majority of interesting formulae, especially when the idea of *some* is present, involve more than one variable. For example, the consideration of the pairs of numbers x and y (fractional or integral) which satisfy $x+y=1$ involves the idea of two correlated variables, x and y. When two variables are present the same two main types of statement occur. For example, 1] for *any* pair of numbers, x and y, $x+y=y+x$, and 2] for *some* pairs of numbers, x and y, $x+y=1$.

The second type of statement invites consideration of the aggregate of pairs of numbers which are bound together by some fixed relation—in the case given, by the relation $x+y=1$. One use of formulae of the first type, true for *any* pair of numbers, is that by them formulae of the second type can be thrown into an indefinite number of equivalent forms. For example, the relation $x+y=1$ is equivalent to the relations

$$y+x=1, \ (x-y)+2y=1, \ 6x+6y=6,$$

and so on. Thus a skilful mathematician uses that equivalent form of the relation under consideration which is most convenient for his immediate purpose.

It is not in general true that, when a pair of terms satisfy some fixed relation, if one of the terms is given the other is also definitely determined. For example, when x and y satisfy $y^2 = x$, if $x = 4$, y can be ± 2, thus, for any positive value of x there are alternative values for y. Also in the relation $x + y > 1$, when either x or y is given, an indefinite number of values remain open for the other.

Again there is another important point to be noticed. If we restrict ourselves to positive numbers, integral or fractional, in considering the relation $x + y = 1$, then, if either x or y be greater than 1, there is no positive number which the other can assume so as to satisfy the relation. Thus the "field" of the relation for x is restricted to numbers less than 1, and similarly for the "field" open to y. Again, consider integral numbers only, positive or negative, and take the relation $y^2 = x$, satisfied by pairs of such numbers. Then whatever integral value is given to y, x can assume one corresponding integral value. So the "field" for y is unrestricted among these positive or negative integers. But the "field" for x is restricted in two ways. In the first place x must be positive, and in the second place, since y is to be integral, x must be a perfect square. Accordingly, the "field" of x is restricted to the set of integers 1^2, 2^2, 3^2, 4^2, and so on, i.e., to 1, 4, 9, 16, and so on.

Methods of Application

The way in which the idea of variables satisfying a relation occurs in the applications of mathematics is worth thought, and by devoting some time to it we shall clear up our thoughts on the whole subject.

Let us start with the simplest of examples:—Suppose that building costs $1s$. per cubic foot and that $20s$. make £1. Then in all the complex circumstances which attend the building of a new house, amid all the various sensations and emotions of the owner, the architect, the builder, the workmen, and the onlookers as the house has grown to completion, this fixed correlation is by the law assumed to hold between the cubic content and the cost to the owner, namely that if x be the number of cubic feet, and £y

the cost, then $20y=x$. This correlation of x and y is assumed to be true for the building of any house by any owner. Also, the volume of the house and the cost are not supposed to have been perceived or apprehended by any particular sensation or faculty, or by any particular man. They are stated in an abstract general way, with complete indifference to the owner's state of mind when he has to pay the bill.

Now think a bit further as to what all this means. The building of a house is a complicated set of circumstances. It is impossible to begin to apply the law, or to test it, unless amid the general course of events it is possible to recognize a definite set of occurrences as forming a particular instance of the building of a house. In short, we must know a house when we see it, and must recognize the events which belong to its building. Then amidst these events, thus isolated in idea from the rest of nature, the two elements of the cost and cubic content must be determinable; and when they are both determined, if the law be true, they satisfy the general formula

$$20y=x.$$

But is the law true? Anyone who has had much to do with building will know that we have here put the cost rather high. It is only for an expensive type of house that it will work out at this price. This brings out another point which must be made clear. While we are making mathematical calculations connected with the formula $20y=x$, it is indifferent to us whether the law be true or false. In fact, the very meanings assigned to x and y, as being a number of cubic feet and a number of pounds sterling, are indifferent. During the mathematical investigation we are, in fact, merely considering the properties of this correlation between a pair of variable numbers x and y. Our results will apply equally well, if we interpret y to mean a number of fishermen and x the number of fish caught, so that the assumed law is that on the average each fisherman catches twenty fish. The mathematical certainty of the investigation only attaches to the results considered as giving properties

of the correlation $20y=x$ between the variable pair of numbers x and y. There is no mathematical certainty whatever
about the cost of the actual building of any house. The law
is not quite true and the result it gives will not be quite
accurate. In fact, it may well be hopelessly wrong.

Now all this no doubt seems very obvious. But in truth
with more complicated instances there is no more common
error than to assume that, because prolonged and accurate
mathematical calculations have been made, the application
of the result to some fact of nature is absolutely certain. The
conclusion of no argument can be more certain than the
assumptions from which it starts. All mathematical calculations about the course of nature must start from some
assumed law of nature, such, for instance, as the assumed
law of the cost of building stated above. Accordingly,
however accurately we have calculated that some event
must occur, the doubt always remains—Is the law true?
If the law states a precise result, almost certainly it is not
precisely accurate; and thus even at the best the result,
precisely as calculated, is not likely to occur. But then we
have no faculty capable of observation with ideal precision,
so, after all, our inaccurate laws may be good enough.

We will now turn to an actual case, that of Newton
and the Law of Gravity. This law states that any two
bodies attract one another with a force proportional to the
product of their masses, and inversely proportional to the
square of the distance between them. Thus if m and M are
the masses of the two bodies, reckoned in lbs. say, and d
miles is the distance between them, the force on either
body, due to the attraction of the other and directed towards
it, is proportional to $\dfrac{mM}{d^2}$; thus this force can be written
as equal to $\dfrac{kmM}{d^2}$, where k is a definite number depending
on the absolute magnitude of this attraction and also on the
scale by which we choose to measure forces. It is easy to see
that, if we wish to reckon in terms of forces such as the
weight of a mass of 1 lb., the number which k represents
must be extremely small; for when m and M and d are each

put equal to 1, $\dfrac{kmM}{d^2}$ becomes the gravitational attraction of two equal masses of 1 lb. at the distance of one mile, and this is quite inappreciable.

However, we have now got our formula for the force of attraction. If we call this force F, it is $F = k\dfrac{mM}{d^2}$, giving the correlation between the variables F, m, M, and d. We all know the story of how it was found out. Newton, it states, was sitting in an orchard and watched the fall of an apple, and then the law of universal gravitation burst upon his mind. It may be that the final formulation of the law occurred to him in an orchard, as well as elsewhere—and he must have been somewhere. But for our purposes it is more instructive to dwell upon the vast amount of preparatory thought, the product of many minds and many centuries, which was necessary before his exact law could be formulated. In the first place, the mathematical habit of mind and the mathematical procedure explained in the previous two chapters had to be generated; otherwise Newton could never have thought of a formula representing the force between *any* two masses at *any* distance. Again, what are the meanings of the terms employed, Force, Mass, Distance? Take the easiest of these terms, Distance. It seems very obvious to us to conceive all material things as forming a definite geometrical whole, such that the distances of the various parts are measurable in terms of some unit length, such as a mile or a yard. This is almost the first aspect of a material structure which occurs to us. It is the gradual outcome of the study of geometry and of the theory of measurement. Even now, in certain cases, other modes of thought are convenient. In a mountainous country distances are often reckoned in hours. But leaving distance, the other terms, Force and Mass, are much more obscure. The exact comprehension of the ideas which Newton meant to convey by these words was of slow growth, and, indeed, Newton himself was the first man who had thoroughly mastered the true general principles of Dynamics.

Throughout the middle ages, under the influence of

Aristotle, the science was entirely misconceived. Newton had the advantage of coming after a series of great men, notably Galileo, in Italy, who in the previous two centuries had reconstructed the science and had invented the right way of thinking about it. He completed their work. Then, finally, having the ideas of force, mass, and distance, clear and distinct in his mind, and realizing their importance and their relevance to the fall of an apple and the motions of the planets, he hit upon the law of gravitation and proved it to be the formula always satisfied in these various motions.

The vital point in the application of mathematical formulae is to have clear ideas and a correct estimate of their relevance to the phenomena under observation. No less than ourselves, our remote ancestors were impressed with the importance of natural phenomena and with the desirability of taking energetic measures to regulate the sequence of events. Under the influence of irrelevant ideas they executed elaborate religious ceremonies to aid the birth of the new moon, and performed sacrifices to save the sun during the crisis of an eclipse. There is no reason to believe that they were more stupid than we are. But at that epoch there had not been opportunity for the slow accumulation of clear and relevant ideas.

A Course of Pure Mathematics is not only the title of G. H. Hardy's most important book, but is also a description of his career. Born in England in 1877, he studied mathematics at Cambridge and spent the greater part of his life teaching there and at Oxford. His primary contributions to mathematics were in the theory of numbers, and he took pride in the "purity" of his mathematical research. In the eyes of the public, his serious work has been dwarfed by a little book, hardly more than an expanded lecture, entitled A Mathematician's Apology, which appeared in 1940. Of it, C. P. Snow, in an address before the American Association for the Advancement of Science, had this to say, "Anyone who has ever worked in any science knows how much aesthetic joy he has obtained. . . . The literature of scientific discovery is full of this aesthetic joy. The very best communication of it that I know is G. H. Hardy's book, A Mathematician's Apology. Graham Greene once said he thought that, along with Henry James's prefaces, this was the best account of the artistic experience ever written."

Yet A Mathematician's Apology has provoked bitter criticism as well as lavish praise. It was Hardy's contempt for applied mathematics ("very little of mathematics is useful practically, and that little is comparatively dull") which caused the British chemist Frederick Soddy to remark in a scornful review which appeared in Nature, "From such cloistral clowning the world sickens." Indeed Hardy himself would have been hard put to it, four years after his book was published, to defend his statement that "real mathematics has no effects on war," in view of his position, also stated in the book, that Einstein, discoverer of the equation $E = mc^2$, was a "real" mathematician.

Nor was this the only basis on which the book was criticised. "What," asked Hardy, "is the proper justification of a mathematician's life?" His answer was that "if a man has a genuine talent, he should be ready to make almost any sacrifice in order to cultivate it to the full." To which Sir Edmund Whittaker, also a great mathematician, asked, with Hardy's other critics, "If Hardy had been conscious of great agility in climbing drainpipes, ought he to have become a cat-burglar? If he had found that he had a great attraction for the other sex, ought he to have set up as a Don Juan?"

To all such animadversions, Hardy would doubtless have replied with ironic laughter. His outrageous pronouncements were part of a personality which, like George Bernard Shaw's, was no doubt exhibitionistic. He dressed eccentrically. He was bellicose in his pacifism. His special hatred was religion in any form. He wrote, "I have never done anything 'useful.' . . . Judged by all practical standards, the value of my mathematical life is nil." Yet in the final paragraph of his book, he says, "The case for my life . . . is this: that I have added something to knowledge and helped others to add more; and

that these somethings have a value which differs in degree
only, and not in kind, from that of the creations of the great
mathematicians, or of any of the other artists, great or small,
who have left some kind of memorial behind them."

A MATHEMATICIAN'S APOLOGY

G. H. HARDY

THE MATHEMATICIAN's patterns, like the painter's or
the poet's, must be *beautiful;* the ideas, like the colors or the
words, must fit together in a harmonious way. Beauty is the
first test: there is no permanent place in the world for ugly
mathematics. And here I must deal with a misconception
which is still widepread (though probably much less so
now than it was twenty years ago), what Whitehead has
called the "literary superstition" that love of and aesthetic
appreciation of mathematics is "a monomania confined to a
few eccentrics in each generation."

The fact is that there are few more "popular" subjects
than mathematics. Most people have some appreciation of
mathematics, just as most people can enjoy a pleasant tune;
and there are probably more people really interested in
mathematics than in music. Appearances may suggest the
contrary, but there are easy explanations. Music can be
used to stimulate mass emotion, while mathematics cannot;
and musical incapacity is recognized (no doubt rightly)
as mildly discreditable, whereas most people are so fright-
ened of the name of mathematics that they are ready, quite
unaffectedly, to exaggerate their own mathematical stupid-
ity.

A very little reflection is enough to expose the absurdity
of the "literary superstition." There are masses of chess-
players in every civilized country—in Russia, almost the
whole educated population; and every chess-player can

recognize and appreciate a "beautiful" game or problem. Yet a chess problem is *simply* an exercise in mathematics (a game not entirely, since psychology also plays a part), and everyone who calls a problem "beautiful" is applauding mathematical beauty, even if it is beauty of a comparatively lowly kind. Chess problems are the hymn-tunes of mathematics.

We may learn the same lesson, at a lower level but for a wider public, from bridge, or descending further, from the puzzle columns of the popular newspapers. Nearly all their immense popularity is a tribute to the drawing power of rudimentary mathematics, and the better makers of puzzles, such as Dudeney or "Caliban," use very little else. They know their business; what the public wants is a little intellectual "kick," and nothing else has quite the kick of mathematics.

I might add that there is nothing in the world which pleases even famous men (and men who have used disparaging language about mathematics) quite so much as to discover, or rediscover, a genuine mathematical theorem. Herbert Spencer republished in his autobiography a theorem about circles which he proved when he was twenty (not knowing that it had been proved over two thousand years before by Plato). Professor Soddy is a more recent and a more striking example (but *his* theorem really is his own).

A chess problem is genuine mathematics, but it is in some way "trivial" mathematics. However ingenious and intricate, however original and surprising the moves, there is something essential lacking. Chess problems are *unimportant*. The best mathematics is *serious* as well as beautiful— "important" if you like, but the word is very ambiguous, and "serious" expresses what I mean much better.

I am not thinking of the "practical" consequences of mathematics. If a chess problem is, in the crude sense, "useless," then that is equally true of most of the best mathematics; that very little of mathematics is useful practically, and that that little is comparatively dull. The "seriousness" of a mathematical theorem lies, not in its practical consequences, which are usually negligible, but in the *significance* of the mathematical ideas which it connects. We may say, roughly, that a mathematical idea is "significant" if it can be connected, in a natural and illuminating way, with a large complex of other mathematical ideas. Thus a serious mathematical theorem, a theorem which connects significant ideas, is likely to lead to important advances in mathematics itself and even in other sciences. No chess problem has ever affected the general development of scientific thought; Pythagoras, Newton, Einstein have in their times changed its whole direction.

The seriousness of a theorem, of course, does not *lie in* its consequences, which are merely the *evidence* for its seriousness. Shakespeare had an enormous influence on the development of the English language, Otway next to none, but that is not why Shakespeare was the better poet. He was the better poet because he wrote much better poetry. The inferiority of the chess problem, like that of Otway's poetry, lies not in its consequences but in its content.

There is one more point which I shall dismiss very shortly, not because it is uninteresting but because it is difficult, and because I have no qualifications for any serious discussion in aesthetics. The beauty of a mathematical theorem *depends* a great deal on its seriousness, as even in poetry the beauty of a line may depend to some extent on the significance of the ideas which it contains. The ideas do matter to the pattern, even in poetry, and much more, naturally, in mathematics; but I must not try to argue the question seriously.

It will be clear by now that, if we are to have any chance of making progress, I must produce examples of "real" mathematical theorems, theorems which every mathematician will admit to be first-rate. And here I am very heavily handicapped by the restrictions under which I am writing. On the one hand my examples must be very simple, and intelligible to a reader who has no specialized mathematical knowledge; no elaborate preliminary explanations must be needed; and a reader must be able to follow the proofs as well as the enunciations. These conditions exclude, for instance, many of the most beautiful theorems of the theory of numbers, such as Fermat's "two square" theorem or the law of quadratic reciprocity. And on the other hand my examples should be drawn from "pukka" mathematics, the mathematics of the working professional mathematician; and this condition excludes a good deal which it would be comparatively easy to make intelligible but which trespasses on logic and mathematical philosophy.

I can hardly do better than go back to the Greeks. I will state and prove two of the famous theorems of Greek mathematics. They are "simple" theorems, simple both in idea and in execution, but there is no doubt at all about their being theorems of the highest class. Each is as fresh and significant as when it was discovered—two thousand years have not written a wrinkle on either of them. Finally, both the statements and the proofs can be mastered in an hour by any intelligent reader, however slender his mathematical equipment.

1]   The first is Euclid's* proof of the existence of an infinity of prime numbers.

The *prime numbers* or *primes* are the numbers

A]   2, 3, 5, 7, 11, 13, 17, 19, 23, 29 . . .

which cannot be resolved into smaller factors.† Thus 37 and 317 are prime. The primes are the material out of which all numbers are built up by multiplication: thus 666=2.3.3.37. Every number which is not prime itself is divisible by at least one prime (usually, of course, by several). We have to prove that there are infinitely many primes, i.e., that the series A] never comes to an end.

Let us suppose that it does, and that

$$2, 3, 5, \ldots , P$$

Elements IX 20. The real origin of many theorems in the *Elements* is obscure, but there seems to be no particular reason for supposing that this one is not Euclid's own.

†There are technical reasons for not counting 1 as a prime.

is the complete series (so that P is the largest prime); and let us, on this hypothesis, consider the number

$$Q= (2.3.5. \ldots .P)+1.$$

It is plain that Q is not divisible by any of 2, 3, 5, $\ldots$, P; for it leaves the remainder 1 when divided by any one of these numbers. But, if not itself prime, it is divisible by *some* prime, and therefore there is a prime (which may be Q itself) greater than any of them. This contradicts our hypothesis, that there is no prime greater than P; and therefore this hypothesis is false.

The proof is by *reductio ad absurdum*, and *reductio ad absurdum*, which Euclid loved so much, is one of a mathematician's finest weapons.* It is a far finer gambit than any chess gambit: a chess player may offer the sacrifice of a pawn or even a piece, but a mathematician offers *the game*.

2] My second example is Pythagoras's† proof of the "irrationality" of $\sqrt{2}$.

A "rational number" is a fraction $\dfrac{a}{b}$, where a and b are integers; we may suppose that a and b have no common factor, since if they had we could remove it. To say that "$\sqrt{2}$ is irrational" is merely another way of saying that 2 cannot be expressed in the form $\left(\dfrac{a}{b}\right)^2$; and this is the same thing as saying that the equation

B] $$a^2=2b^2$$

cannot be satisfied by integral values of a and b which have no common factor. This is a theorem of pure arithmetic, which does not demand any knowledge of "irrational numbers" or depend on any theory about their nature.

We argue again by *reductio ad absurdum*; we suppose

---

*The proof can be arranged so as to avoid a *reductio,* and logicians of some schools would prefer that it should be.
†The proof traditionally ascribed to Pythagoras, and certainly a product of his school. The theorem occurs, in a much more general form, in Euclid (*Elements* x 9).

that B] is true, a and b being integers without any common factor. It follows from B] that a^2 is even (since $2b^2$ is divisible by 2), and therefore that a is even (since the square of an odd number is odd). If a is even then

C] $$a = 2c$$

for some integral value of c; and therefore

$$2b^2 = a^2 = (2c)^2 = 4c^2$$

or

D] $$b^2 = 2c^2.$$

Hence b^2 is even, and therefore (for the same reason as before) b is even. That is to say, a and b are both even, and so have the common factor 2. This contradicts our hypothesis, and therefore is false.

It follows from Pythagoras's theorem that the diagonal of a square is incommensurable with the side (that their ratio is not a rational number, that there is no unit of which both are integral multiples). For if we take the side as our unit of length, and the length of the diagonal is d, then, by a very familiar theorem also ascribed to Pythagoras,[*]

$$d^2 = 1^2 + 1^2 = 2,$$

so that d cannot be a rational number.

I could quote any number of fine theorems from the theory of numbers whose *meaning* anyone can understand. For example, there is what is called "the fundamental theorem of arithmetic," that any integer can be resolved, *in one way only*, into a product of primes. Thus $666 = 2.3.3.37$, and there is no other decomposition; it is impossible that $666 = 2.11.29$ or that $13.89 = 17.73$ (and we can see so without working out the products). This theorem is, as its name implies, the foundation of higher arithmetic; but the proof, although not "difficult," requires a certain amount of preface and might be found tedious by an unmathematical reader.

Another famous and beautiful theorem is Fermat's "two

[*]Euclid, *Elements* I 47.

square" theorem. The primes may (if we ignore the special prime 2) be arranged in two classes; the primes

$$5, 13, 17, 29, 37, 41 \ldots$$

which leave remainder 1 when divided by 4, and the primes

$$3, 7, 11, 19, 23, 31 \ldots$$

which leave remainder 3. All the primes of the first class, and none of the second, can be expressed as the sum of two integral squares: thus

$$5 = 1^2 + 2^2, \quad 13 = 2^2 + 3^2,$$
$$17 = 1^2 + 4^2, \quad 29 = 2^2 + 5^2;$$

but 3, 7, 11, and 19 are not expressible in this way (as the reader may check by trial). This is Fermat's theorem, which is ranked, very justly, as one of the finest of arithmetic. Unfortunately there is no proof within the comprehension of anybody but a fairly expert mathematician.

There are also beautiful theorems in the "theory of aggregates" (*Mengenlehre*), such as Cantor's theorem of the "nonenumerability" of the continuum. Here there is just the opposite difficulty. The proof is easy enough, when once the language has been mastered, but considerable explanation is necessary before the *meaning* of the theorem becomes clear. So I will not try to give more examples. Those which I have given are test cases, and a reader who cannot appreciate them is unlikely to appreciate anything in mathematics.

I said that a mathematician was a maker of patterns of ideas, and that beauty and seriousness were the criteria by which his patterns should be judged. I can hardly believe that anyone who has understood the two theorems will dispute that they pass these tests. If we compare them with Dudeney's most ingenious puzzles, or the finest chess problems that masters of that art have composed, their superiority in both respects stands out: there is an unmistakable difference of class. They are much more serious, and also much more beautiful; can we define, a little more closely where their superiority lies?

In the first place, the superiority of the mathematical theorems in *seriousness* is obvious and overwhelming. The chess problem is the product of an ingenious but very limited complex of ideas, which do not differ from one another very fundamentally and have no external repercussions. We should think in the same way if chess had never been invented, whereas the theorems of Euclid and Pythagoras have influenced thought profoundly, even outside mathematics.

Thus Euclid's theorem is vital for the whole structure of arithmetic. The primes are the raw material out of which we have to build arithmetic, and Euclid's theorem assures us that we have plenty of material for the task. But the theorem of Pythagoras has wider applications and provides a better text.

We should observe first that Pythagoras's argument is capable of far-reaching extension, and can be applied, with little change of principle, to very wide classes of "irrationals." We can prove very similarly (as Theaetetus seems to have done) that

$$\sqrt{3}, \ \sqrt{5}, \ \sqrt{7}, \ \sqrt{11}, \ \sqrt{13}, \ \sqrt{17}$$

are irrational, or (going beyond Theaetetus) that $\sqrt[3]{2}$ and $\sqrt[3]{17}$ are irrational.*

Euclid's theorem tells us that we have a good supply of material for the construction of a coherent arithmetic of the integers. Pythagoras's theorem and its extensions tell us that, when we have constructed this arithmetic, it will not prove sufficient for our needs, since there will be many magnitudes which obtrude themselves upon our attention and which it will be unable to measure; the diagonal of the square is merely the most obvious example. The profound importance of this discovery was recognized at once by the Greek mathematicians. They had begun by assuming (in accordance, I suppose, with the "natural" dictates of "common sense") that all magnitudes of the same kind are

---

*See Ch. IV of Hardy and Wright's *Introduction to the Theory of Numbers*, where there are discussions of different generalizations of Pythagoras's argument, and of a historical puzzle about Theaetetus.

commensurable, that any two lengths, for example, are multiples of some common unit, and they had constructed a theory of proportion based on this assumption. Pythagoras's discovery exposed the unsoundness of this foundation, and led to the construction of the much more profound theory of Eudoxus which is set out in the fifth book of the *Elements*, and which is regarded by many modern mathematicians as the finest achievement of Greek mathematics. This theory is astonishingly modern in spirit, and may be regarded as the beginning of the modern theory of irrational number, which has revolutionized mathematical analysis and had much influence on recent philosophy.

There is no doubt at all, then, of the "seriousness" of either theorem. It is therefore the better worth remarking that neither theorem has the slightest "practical" importance. In practical applications we are concerned only with comparatively small numbers; only stellar astronomy and atomic physics deal with "large" numbers, and they have very little more practical importance, as yet, than the most abstract pure mathematics. I do not know what is the highest degree of accuracy which is ever useful to an engineer—we shall be very generous if we say ten significant figures. Then

$$3 \cdot 14159265$$

(the value of π to eight places of decimals) is the ratio

$$\frac{314159265}{1000000000}$$

of two numbers of ten digits. The number of primes less than 1,000,000,000 is 50,847,478: that is enough for an engineer, and he can be perfectly happy without the rest. So much for Euclid's theorem; and, as regards Pythagoras's, it is obvious that irrationals are uninteresting to an engineer, since he is concerned only with approximations, and all approximations are rational.

A "serious" theorem is a theorem which contains "significant" ideas, and I suppose that I ought to try to analyze a little more closely the qualities which make a mathematical idea significant. This is very difficult, and it is unlikely that any analysis which I can give will be very valuable. We can recognize a "significant" idea when we see it, as we can those which occur in my two standard theorems; but this power of recognition requires a rather high degree of mathematical sophistication, and of that familiarity with mathematical ideas which comes only from many years spent in their company. So I must attempt some sort of analysis; and it should be possible to make one which, however inadequate, is sound and intelligible so far as it goes. There are two things at any rate which seem essential, a certain *generality* and a certain *depth;* but neither quality is easy to define at all precisely.

A significant mathematical idea, a serious mathematical theorem, should be "general" in some such sense as this. The idea should be one which is a constituent in many mathematical constructs, which is used in the proof of theorems of many different kinds. The theorem should be one which, even if stated originally (like Pythagoras's theorem) in a quite special form, is capable of considerable extension and is typical of a whole class of theorems of its kind. The relations revealed by the proof should be such as connect many different mathematical ideas. All this is very vague, and subject to many reservations. But it is easy enough to see that a theorem is unlikely to be serious when it lacks these qualities conspicuously; we have only to take examples from the isolated curiosities in which arithmetic abounds. I take two, almost at random, from Rouse Ball's *Mathematical Recreations.**

a]   8712 and 9801 are the only four-figure numbers which are integral multiples of their "reversals":

$$8712 = 4.2178, \quad 9801 = 9.1089,$$

and there are no other numbers below 10,000 which have this property.

b]   There are just four numbers (after 1) which are the sums of the cubes of their digits, viz.

*11th edition, 1939 (revised by H. S. M. Coxeter).

$$153 = 1^3 + 5^3 + 3^3, \quad 370 = 3^3 + 7^3 + 0^3,$$
$$371 = 3^3 + 7^3 + 1^3, \quad 407 = 4^3 + 0^3 + 7^3.$$

These are odd facts, very suitable for puzzle columns and likely to amuse amateurs, but there is nothing in them which appeals much to a mathematician. The proofs are neither difficult nor interesting—merely a little tiresome. The theorems are not serious; and it is plain that one reason (though perhaps not the most important) is the extreme speciality of both the enunciations and the proofs, which are not capable of any significant generalization.

"Generality" is an ambiguous and rather dangerous word, and we must be careful not to allow it to dominate our discussion too much. It is used in various senses both in mathematics and in writings about mathematics, and there is one of these in particular, on which logicians have very properly laid great stress, which is entirely irrelevant here. In this sense, which is quite easy to define, *all* mathematical theorems are equally and completely "general."

"The certainty of mathematics," says Whitehead,* "depends on its complete abstract generality." When we assert that $2+3 = 5$, we are asserting a relation between three groups of "things"; and these "things" are not apples or pennies, or things of any one particular sort or another, but *just* things, "any old things." The meaning of the statement is entirely independent of the individualities of the members of the groups. All mathematical "objects" or "entities" or "relations," such as "2," "3," "5," "+," or "=," and all mathematical propositions in which they occur, are completely general in the sense of being completely abstract. Indeed one of Whitehead's words is superfluous, since generality, in his sense, *is* abstractness.

This sense of the word is important, and the logicians are quite right to stress it, since it embodies a truism which a good many people who ought to know better are apt to forget. It is quite common, for example, for an astronomer or a physicist to claim that he has found a "mathematical

*Science and the Modern World, p. 33.

proof" that the physical universe must behave in a particular way. All such claims, if interpreted literally, are strictly nonsense. It *cannot* be possible to prove mathematically that there will be an eclipse tomorrow, because eclipses, and other physical phenomena, do not form part of the abstract world of mathematics; and this, I suppose, all astronomers would admit when pressed, however many eclipses they may have predicted correctly.

It is obvious that we are not concerned with this sort of "generality" now. We are looking for *differences* of generality between one mathematical theorem and another, and in Whitehead's sense all are equally general. Thus the "trivial theorems a] and b] on pages 122 and 123 are just as "abstract" or "general" as those of Euclid and Pythagoras, and so is a chess problem. It makes no difference to a chess problem whether the pieces are white and black, or red and green, or whether there are physical "pieces" at all; it is the *same* problem which an expert carries easily in his head and which we have to reconstruct laboriously with the aid of the board. The board and the pieces are mere devices to stimulate our sluggish imaginations, and are no more essential to the problem than the blackboard and the chalk are to the theorems in a mathematical lecture.

It is not this kind of generality, common to all mathematical theorems, which we are looking for now, but the more subtle and elusive kind of generality which I tried to describe in rough terms. And we must be careful not to lay *too* much stress even on generality of this kind (as I think logicians like Whitehead tend to do). It is not mere "piling of subtlety of generalization upon subtlety of generalization"* which is the outstanding achievement of modern mathematics. Some measure of generality must be present in any high-class theorem, but *too much* tends inevitably to insipidity. "Everything is what it is, and not another thing," and the differences between things are quite as interesting as their resemblances. We do not choose our friends because they embody all the pleasant qualities of humanity, but because they are the people that they are. And so in

Science and the Modern World, p. 44.

mathematics; a property common to too many objects can hardly be very exciting, and mathematical ideas also become dim unless they have plenty of individuality. Here at any rate I can quote Whitehead on my side: "it is the large generalization, limited by a happy particularity, which is the fruitful conception." *

The second quality which I demanded in a significant idea was *depth*, and this is still more difficult to define. It has *something* to do with *difficulty*; the "deeper" ideas are usually the harder to grasp: but it is not at all the same. The ideas underlying Pythagoras's theorem and its generalizations are quite deep, but no mathematician now would find them difficult. On the other hand a theorem may be essentially superficial and yet quite difficult to prove (as are many "Diophantine" theorems, i.e., theorems about the solution of equations in integers).

It seems that mathematical ideas are arranged somehow in strata, the ideas in each stratum being linked by a complex of relations both among themselves and with those above and below. The lower the stratum, the deeper (and in general the more difficult) the idea. Thus the idea of an "irrational" is deeper than that of an integer; and Pythagoras's theorem is, for that reason, deeper than Euclid's.

Let us concentrate our attention on the relations between the integers, or some other group of objects lying in some particular stratum. Then it may happen that one of these relations can be comprehended completely, that we can recognize and prove, for example, some property of the integers, without any knowledge of the contents of lower strata. Thus we proved Euclid's theorem by consideration of properties of integers only. But there are also many theorems about integers which we cannot appreciate properly, and still less prove, without digging deeper and considering what happens below.

It is easy to find examples in the theory of prime numbers. Euclid's theorem is very important, but not very deep: we can prove that there are infinitely many primes without

Science and the Modern World, p. 46.

using any notion deeper than that of "divisibility." But new questions suggest themselves as soon as we know the answer to this one. There is an infinity of primes, but how is this infinity distributed? Given a large number N, say 10^{80} or $10^{10^{10}}$,* about how many primes are there less than N?† When we ask *these* questions, we find ourselves in a quite different position. We can answer them, with rather surprising accuracy, but only by boring much deeper, leaving the integers above us for a while, and using the most powerful weapons of the modern theory of functions. Thus the theorem which answers our questions (the so-called "Prime Number Theorem") is a much deeper theorem than Euclid's or even Pythagoras's.

I could multiply examples, but this notion of "depth" is an elusive one even for a mathematician who can recognize it, and I can hardly suppose that I could say anything more about it here which would be of much help to other readers.

*It is supposed that the number of protons in the universe is about 10^{80}. The number $10^{10^{10}}$, if written at length, would occupy about 50,000 volumes of average size.
†As I mentioned, there are 50,847,478 primes less than 1,000,-000,000; but that is as far as our *exact* knowledge extends.

In a widely quoted passage, Bertrand Russell states, "Mathematics, rightly viewed, possesses not only truth but supreme beauty—a beauty cold and austere . . . yet sublimely pure. . . . Remote from human passions, remote even from the pitiful facts of nature, the generations have gradually created an ordered cosmos, where pure thought can dwell as in its natural home. . . ." Something of the same idea is expressed in the first paragraph of the following article by Henri Poincaré.

The problem of mathematical discovery is part of a larger one—that of scientific discovery, and that in turn is part of one still larger—the genesis of all creative activity. It has been examined not only by psychologists and critics, but more significantly by creative workers themselves. Louis Pasteur epitomized one aspect of the subject when he wrote that "chance favors only the prepared mind." Walter Bradford Cannon, the physiologist, elaborated on it in "Gains From Serendipity," and since the publication of the book in which his article appeared, the synonym for accident and luck which he borrowed from Horace Walpole has become part of the common vocabulary. A single well-known example of such serendipity was Fleming's discovery of penicillin, but this example can be multiplied manyfold.

Frequently, however, in the process known as "inspiration," chance plays little or no part. In Imagination Creatrix, John Livingston Lowes, the Harvard scholar, described the road which led Coleridge to "The Ancient Mariner." It was a long storing-up of knowledge and impressions which exploded into "the flash of amazing vision" and then into "the exacting task of translating the vision into actuality." The legendary fall of the apple, says Lowes, triggered Newton's cosmic conception. A joyful moment of vision revealed the theory of evolution to Darwin as he was riding one day in his carriage. And these examples from art and science have their parallel in the moment of inspiration granted to Poincaré in his study of Fuchsian functions.

Jules Henri Poincaré was born at Nancy, France, in 1854 and died in Paris in 1912. From 1881, he was Professor of Mathematics in the Faculty of Sciences at Paris. His work on the theory of functions, which he here describes, had profound effects on the application of mathematics to physics and the measurement of mechanical systems. In the breadth of his mathematical knowledge he was unsurpassed, and his fertility was astounding. He was the author of hundreds of papers and was recognized during his lifetime as one of the immortals of mathematics. His extraordinary erudition was matched by the felicity of his prose style, and by the charm of his personality, which made him one of the most beloved Frenchmen of his generation.

MATHEMATICAL DISCOVERY

HENRI POINCARÉ

THE GENESIS of mathematical discovery is a problem which must inspire the psychologist with the keenest interest. For this is the process in which the human mind seems to borrow least from the exterior world, in which it acts, or appears to act, only by itself and on itself, so that by studying the process of geometric thought we may hope to arrive at what is most essential in the human mind.

One first fact must astonish us, or rather would astonish us if we were not too much accustomed to it. How does it happen that there are people who do not understand mathematics? If the science invokes only the rules of logic, those accepted by all well-formed minds, if its evidence is founded on principles that are common to all men, and that none but a madman would attempt to deny, how does it happen that there are so many people who are entirely impervious to it?

There is nothing mysterious in the fact that every one is not capable of discovery. That every one should not be able to retain a demonstration he has once learnt is still comprehensible. But what does seem most surprising, when we consider it, is that any one should be unable to understand a mathematical argument at the very moment it is stated to him. And yet those who can only follow the argument with difficulty are in a majority; this is incontestable, and the experience of teachers of secondary education will certainly not contradict me.

And still further, how is error possible in mathematics? A healthy intellect should not be guilty of any error in logic, and yet there are very keen minds which will not make a false step in a short argument such as those we have to make in the ordinary actions of life, which yet are incapable

128

of following or repeating without error the demonstrations of mathematics which are longer, but which are, after all, only accumulations of short arguments exactly analogous to those they make so easily. Is it necessary to add that mathematicians themselves are not infallible?

The answer appears to me obvious. Imagine a long series of syllogisms in which the conclusions of those that precede form the premises of those that follow. We shall be capable of grasping each of the syllogisms, and it is not in the passage from premises to conclusion that we are in danger of going astray. But between the moment when we meet a proposition for the first time as the conclusion of one syllogism, and the moment when we find it once more as the premise of another syllogism, much time will sometimes have elapsed, and we shall have unfolded many links of the chain; accordingly it may well happen that we shall have forgotten it, or, what is more serious, forgotten its meaning. So we may chance to replace it by a somewhat different proposition, or to preserve the same statement but give it a slightly different meaning, and thus we are in danger of falling into error.

A mathematician must often use a rule, and, naturally, he begins by demonstrating the rule. At the moment the demonstration is quite fresh in his memory he understands perfectly its meaning and significance, and he is in no danger of changing it. But later on he commits it to memory, and only applies it in a mechanical way, and then, if his memory fails him, he may apply it wrongly. It is thus, to take a simple and almost vulgar example, that we sometimes make mistakes in calculation, because we have forgotten our multiplication table.

On this view special aptitude for mathematics would be due to nothing but a very certain memory or a tremendous power of attention. It would be a quality analogous to that of the whist player who can remember the cards played, or, to rise a step higher, to that of the chess player who can picture a very great number of combinations and retain them in his memory. Every good mathematician should also be a good chess player and *vice versa*, and similarly he should be a good numerical calculator. Certainly this some-times happens, and thus Gauss was at once a geometrician of genius and a very precocious and very certain calculator.

But there are exceptions, or rather I am wrong, for I cannot call them exceptions, otherwise the exceptions would be more numerous than the cases of conformity with the rule. On the contrary, it was Gauss who was an exception. As for myself, I must confess I am absolutely incapable of doing an addition sum without a mistake. Similarly I should be a very bad chess player. I could easily calculate that by playing in a certain way I should be exposed to such and such a danger; I should then review many other moves, which I should reject for other reasons, and I should end by making the move I first examined, having forgotten in the interval the danger I had foreseen.

In a word, my memory is not bad, but it would be in-sufficient to make me a good chess player. Why, then, does it not fail me in a difficult mathematical argument in which the majority of chess players would be lost? Clearly because it is guided by the general trend of the argument. A mathe-matical demonstration is not a simple juxtaposition of syl-logisms; it consists of syllogisms placed in a certain order, and the order in which these elements are placed is much more important than the elements themselves. If I have the feeling, so to speak the intuition, of this order, so that I can perceive the whole of the argument at a glance, I need no longer be afraid of forgetting one of the elements; each of them will place itself naturally in the position prepared for it, without my having to make any effort of memory.

It seems to me, then, as I repeat an argument I have learnt, that I could have discovered it. This is often only an illusion; but even then, even if I am not clever enough to create for myself, I rediscover it myself as I repeat it.

We can understand that this feeling, this intuition of mathematical order, which enables us to guess hidden harmonies and relations, cannot belong to every one. Some have neither this delicate feeling that is difficult to define, nor a power of memory and attention above the common, and so they are absolutely incapable of understanding even the first steps of higher mathematics. This applies to the majority of people. Others have the feeling only in a slight degree, but they are gifted with an uncommon memory and a great capacity for attention. They learn the details one after the other by heart, they can understand mathematics and sometimes apply them, but they are not in a condition to create. Lastly, others possess the special intuition I have spoken of more or less highly developed, and they can not only understand mathematics, even though their memory is in no way extraordinary, but they can become creators, and seek to make discovery with more or less chance of success, according as their intuition is more or less developed.

What, in fact, is mathematical discovery? It does not consist in making new combinations with mathematical entities that are already known. That can be done by any one, and the combinations that could be so formed would be infinite in number, and the greater part of them would be absolutely devoid of interest. Discovery consists precisely in not constructing useless combinations, but in constructing those that are useful, which are an infinitely small minority. Discovery is discernment, selection.

How this selection is to be made I have explained above. Mathematical facts worthy of being studied are those which, by their analogy with other facts, are capable of conducting us to the knowledge of a mathematical law, in the same way that experimental facts conduct us to the knowledge of a physical law. They are those which reveal unsuspected relations between other facts, long since known, but wrongly believed to be unrelated to each other.

Among the combinations we choose, the most fruitful are often those which are formed of elements borrowed from widely separated domains. I do not mean to say that for discovery it is sufficient to bring together objects that are as incongruous as possible. The greater part of the combinations so formed would be entirely fruitless, but some among them, though very rare, are the most fruitful of all.

Discovery, as I have said, is selection. But this is perhaps not quite the right word. It suggests a purchaser who has been shown a large number of samples, and examines them one after the other in order to make his selection. In our case the samples would be so numerous that a whole life would not give sufficient time to examine them. Things do not happen in this way. Unfruitful combinations do not so much as present themselves to the mind of the discoverer. In the field of his consciousness there never appear any but really useful combinations, and some that he rejects, which, however, partake to some extent of the character of useful combinations. Everything happens as if the discoverer were a secondary examiner who had only to interrogate candidates declared eligible after passing a preliminary test.

But what I have said up to now is only what can be observed or inferred by reading the works of geometricians, provided they are read with some reflection.

It is time to penetrate further, and to see what happens in the very soul of the mathematician. For this purpose I think I cannot do better than recount my personal recollections. Only I am going to confine myself to relating how I wrote my first treatise on Fuchsian functions. I must apologize, for I am going to introduce some technical expressions, but they need not alarm the reader, for he has no need to understand them. I shall say, for instance, that I found the demonstration of such and such a theorem under such and such circumstances; the theorem will have a barbarous name that many will not know, but that is of no importance. What is interesting for the psychologist is not the theorem but the circumstances.

For a fortnight I had been attempting to prove that there could not be any function analogous to what I have since called Fuchsian functions. I was at that time very ignorant. Every day I sat down at my table and spent an hour or two trying a great number of combinations, and I arrived at no result. One night I took some black coffee, contrary to my custom, and was unable to sleep. A host of ideas kept surging in my head; I could almost feel them jostling one another, until two of them coalesced, so to speak, to form a stable combination. When morning came, I had established the existence of one class of Fuchsian functions, those that are derived from the hypergeometric series. I had only to verify the results, which only took a few hours.

Then I wished to represent these functions by the quotient of two series. This idea was perfectly conscious and deliberate; I was guided by the analogy with elliptical functions. I asked myself what must be the properties of these series, if they existed, and I succeeded without difficulty in forming the series that I have called Theta-Fuchsian.

At this moment I left Caen, where I was then living, to take part in a geological conference arranged by the School of Mines. The incidents of the journey made me forget my mathematical work. When we arrived at Coutances, we got into a break to go for a drive, and, just as I put my foot on the step, the idea came to me, though nothing in my former thoughts seemed to have prepared me for it, that the transformations I had used to define Fuchsian functions were identical with those of non-Euclidian geometry. I made no verification, and had no time to do so, since I took up the conversation again as soon as I had sat down in the break, but I felt absolute certainty at once. When I got back to Caen I verified the result at my leisure to satisfy my conscience.

I then began to study arithmetical questions without any great apparent result, and without suspecting that they could have the least connection with my previous researches. Disgusted at my want of success, I went away to spend a few days at the seaside, and thought of entirely different things. One day, as I was walking on the cliff, the idea came to me, again with the same characteristics of conciseness, suddenness, and immediate certainty, that arithmetical transformations of indefinite ternary quadratic forms are identical with those of non-Euclidian geometry.

Returning to Caen, I reflected on this result and deduced its consequences. The example of quadratic forms showed me that there are Fuchsian groups other than those which correspond with the hypergeometric series; I saw that I could apply to them the theory of the Theta-Fuchsian series, and that, consequently, there are Fuchsian functions other than those which are derived from the hypergeometric series, the only ones I knew up to that time. Naturally, I proposed to form all these functions. I laid siege to them systematically and captured all the outworks one after the other. There was one, however, which still held out, whose fall would carry with it that of the central fortress. But all my efforts were of no avail at first, except to make me better understand the difficulty, which was already something. All this work was perfectly conscious.

Thereupon I left for Mont-Valérien, where I had to serve my time in the army, and so my mind was preoccupied with very different matters. One day, as I was crossing the street, the solution of the difficulty which had brought me to a standstill came to me all at once. I did not try to fathom it immediately, and it was only after my service was finished that I returned to the question. I had all the elements, and had only to assemble and arrange them. Accordingly I composed my definitive treatise at a sitting and without any difficulty.

It is useless to multiply examples, and I will content myself with this one alone. As regards my other researches, the accounts I should give would be exactly similar, and the observations related by other mathematicians in the enquiry of *l'Enseignement Mathématique* would only confirm them.

One is at once struck by these appearances of sudden illumination, obvious indications of a long course of previous unconscious work. The part played by this unconscious work in mathematical discovery seems to me indisputable, and we shall find traces of it in other cases where it is less evident. Often when a man is working at a difficult question, he accomplishes nothing the first time he sets to work. Then he takes more or less of a rest, and sits down again at his table. During the first half-hour he still finds nothing, and then all at once the decisive idea presents itself to his mind. We might say that the conscious work proved more fruitful because it was interrupted and the rest restored force and freshness to the mind. But it is more probable that the rest was occupied with unconscious work, and that the result of this work was afterwards revealed to the geometrician exactly as in the cases I have quoted, except that the revelation, instead of coming to light during a walk or a journey, came during a period of conscious work, but independently of that work, which at most only performs the unlocking process, as if it were the spur that excited into conscious form the results already acquired during the rest, which till then remained unconscious.

There is another remark to be made regarding the conditions of this unconscious work, which is, that it is not possible, or in any case not fruitful, unless it is first preceded and then followed by a period of conscious work. These sudden inspirations are never produced (and this is sufficiently proved already by the examples I have quoted) except after some days of voluntary efforts which appeared absolutely fruitless, in which one thought one had accomplished nothing, and seemed to be on a totally wrong track. These efforts, however, were not as barren as one thought; they set the unconscious machine in motion, and without them it would not have worked at all, and would not have produced anything.

Such are the facts of the case, and they suggest the following reflections. The result of all that precedes is to show that the unconscious ego, or, as it is called, the subliminal ego, plays a most important part in mathematical discovery. But the subliminal ego is generally thought of as purely automatic. Now we have seen that mathematical work is not a simple mechanical work, and that it could not be entrusted to any machine, whatever the degree of perfection we suppose it to have been brought to. It is not merely a question of applying certain rules, of manufacturing as many combinations as possible according to certain fixed laws. The combinations so obtained would be extremely numerous, useless, and encumbering. The real work of the discoverer consists in choosing between these combinations with a view to eliminating those that are useless, or rather not giving himself the trouble of making them at all. The rules which must guide this choice are extremely subtle and delicate, and it is practically impossible to state them in precise language; they must be felt rather than formulated.

Under these conditions, how can we imagine a sieve capable of applying them mechanically?

The following, then, presents itself as a first hypothesis. The subliminal ego is in no way inferior to the conscious ego; it is not purely automatic; it is capable of discernment; it has tact and lightness of touch; it can select, and it can divine. More than that, it can divine better than the conscious ego, since it succeeds where the latter fails. In a word, is not the subliminal ego superior to the conscious ego? The importance of this question will be readily understood. In a recent lecture, M. Boutroux showed how it had arisen on entirely different occasions, and what consequences would be involved by an answer in the affirmative. (See also the same author's *Science et Religion*, pp. 313 *et seq.*)

Are we forced to give this affirmative answer by the facts I have just stated? I confess that, for my part, I should be loth to accept it. Let us, then, return to the facts, and see if they do not admit of some other explanation.

It is certain that the combinations which present themselves to the mind in a kind of sudden illumination after a somewhat prolonged period of unconscious work are generally useful and fruitful combinations, which appear to be the result of a preliminary sifting. Does it follow from this that the subliminal ego, having divined by a delicate intuition that these combinations could be useful, has formed none but these, or has it formed a great many others which were devoid of interest, and remained unconscious?

Under this second aspect, all the combinations are formed as a result of the automatic action of the subliminal ego, but those only which are interesting find their way into the field of consciousness. This, too, is most mysterious. How can we explain the fact that, of the thousand products of our unconscious activity, some are invited to cross the threshold, while others remain outside? Is it mere chance that gives them this privilege? Evidently not. For instance, of all the excitements of our senses, it is only the most intense that retain our attention, unless it has been directed upon them by other causes. More commonly the privileged unconscious phenomena, those that are capable of becoming conscious, are those which, directly or indirectly, most deeply affect our sensibility.

It may appear surprising that sensibility should be introduced in connection with mathematical demonstrations, which, it would seem, can only interest the intellect. But not if we bear in mind the feeling of mathematical beauty, of the harmony of numbers and forms and of geometric elegance. It is a real æsthetic feeling that all true mathematicians recognize, and this is truly sensibility.

Now, what are the mathematical entities to which we attribute this character of beauty and elegance, which are capable of developing in us a kind of æsthetic emotion? Those whose elements are harmoniously arranged so that the mind can, without effort, take in the whole without neglecting the details. This harmony is at once a satisfaction to our æsthetic requirements, and an assistance to the mind which it supports and guides. At the same time, by setting before our eyes a well-ordered whole, it gives us a presentiment of a mathematical law. Now, as I have said above, the only mathematical facts worthy of retaining our attention and capable of being useful are those which can make us acquainted with a mathematical law. Accordingly we arrive at the following conclusion. The useful combinations are precisely the most beautiful, I mean those that can most charm that special sensibility that all mathematicians know, but of which laymen are so ignorant that they are often tempted to smile at it.

III. Some Branches of MATHEMATICS

III. Some Branches of Mathematics

In all probability, the most elementary exercise in mathematics is that of counting. The concepts of few, many, and how many have been familiar to most of us from earliest childhood. The mental processes involved seem to require little education or culture—they seem to be almost instinctive with normally intelligent human beings. Actually, as we have mentioned at the beginning of this book, and as we may learn from George Gamow's Hungarian aristocrat (for whom we may substitute the untutored savage), the idea of number is one of considerable sophistication, one which, even among such civilized peoples as the Egyptians and the Greeks, was not thoroughly understood. This idea of number, of counting, which is fundamental to arithmetic, is the subject of the following article by a distinguished and colorful mathematical physicist.

George Gamow was born at Odessa, Russia, in 1904, and was educated there and at the University of Leningrad. He taught at the universities of Copenhagen, Paris, and London, became Professor of Theoretical Physics at George Washington University in 1934, and now occupies a similar post at the University of Colorado. His contributions to physics have been in the nuclear and thermonuclear fields, in cosmology and the origin of the elements. He is also well known for his sound and entertaining popularizations of science, among them Mr. Tompkins in Wonderland, and The Birth and Death of the Sun, the latter an explanation of his own theory.

BIG NUMBERS

GEORGE GAMOW

1. *How High Can You Count?*

THERE IS A STORY about two Hungarian aristocrats who decided to play a game in which the one who calls the largest number wins.

"Well," said one of them, "you name your number first."

After a few minutes of hard mental work the second aristocrat finally named the largest number he could think of.

"Three," he said.

Now it was the turn of the first one to do the thinking, but after a quarter of an hour he finally gave up.

"You've won," he agreed.

Of course these two Hungarian aristocrats do not represent a very high degree of intelligence and this story is probably just a malicious slander, but such a conversation might actually have taken place if the two men had been, not Hungarians, but Hottentots. We have it indeed on the authority of African explorers that many Hottentot tribes do not have in their vocabulary the names for numbers larger than three. Ask a native down there how many sons he has or how many enemies he has slain, and if the number is more than three he will answer "many." Thus in the Hottentot country in the art of counting fierce warriors would be beaten by an American child of kindergarten age who could boast the ability to count up to ten!

Nowadays we are quite accustomed to the idea that we can write as big a number as we please—whether it is to represent war expenditures in cents, or stellar distances in inches—by simply setting down a sufficient number of zeros on the right side of some figure. You can put in zeros until your hand gets tired, and before you know it you will have a number larger than even the total number of atoms in the universe,[1] which, incidentally, is 300,000,000,000,000,-000,000,000,000,000,000,000,000,000,000,000,000,-000,000,000,000,000,000.

Or you may write it in this shorter form: $3 \cdot 10^{74}$.

Here the little number ^{74} above and to the right of 10 indicates that there must be that many zeros written out, or, in other words 3 must be multiplied by 10 seventy-four times.

But this "arithmetic-made-easy" system was not known in ancient times. In fact it was invented less than two thousand years ago by some unknown Indian mathematician. Before his great discovery—and it *was* a great discovery, although we usually do not realize it—numbers were written by using a special symbol for each of what we now call decimal units, and repeating this symbol as many times as there were units. For example the number 8732 was written by ancient Egyptians:

whereas a clerk in Caesar's office would have represented it in this form:

MMMMMMMMDCCXXXII

The latter notations must be familiar to you, since Roman numerals are still used sometimes—to indicate the volumes or chapters of a book, or to give the date of a historical event on a pompous memorial tablet. Since, however, the needs of ancient accounting did not exceed the numbers of a few thousands, the symbols for higher decimal units were nonexistent, and an ancient Roman, no matter how well trained in arithmetic, would have been extremely embarrassed if he had been asked to write "one million." The best he could have done to comply with the request, would have been to write one thousand *M s* in succession, which would have taken many hours of hard work.

1.   Measured as far as the largest telescope can penetrate.

For the ancients, very large numbers such as those of the stars in the sky, the fish in the sea, or grains of sand on the beach were "incalculable," just as for a Hottentot "five" is incalculable, and becomes simply "many"!

It took the great brain of Archimedes, a celebrated scientist of the third century B.C., to show that it is possible to write really big numbers. In his treatise *The Psammites*, or *Sand Reckoner*, Archimedes says:

There are some who think that the number of sand grains is infinite in multitude; and I mean by sand not only that which exists about Syracuse and the rest of Sicily, but all the grains of sand which may be found in all the regions of the Earth, whether inhabited or uninhabited. Again there are some who, without regarding the number as infinite, yet think that *no number can be named which is great enough to exceed that which would designate the number of the Earth's grains of sand*. And it is clear that those who hold this view, if they imagined a mass made up of sand in other respects as large as the mass of the Earth, including in it all the seas and all the hollows of the Earth filled up to the height of the highest mountains, would be still more certain that no number could be expressed which would be larger than that needed to represent the grains of sand thus accumulated. But I will try to show that of the numbers named by me some exceed not only the number of grains of sand which would make a mass equal in size to the Earth filled up in the way described, but even equal to a mass the size of the Universe.

The way to write very large numbers proposed by Archimedes in this famous work is similar to the way large numbers are written in modern science. He begins with the largest number that existed in ancient Greek arithmetic: a "myriad," or ten thousand. Then he introduced a new number, "a myriad myriad" (a hundred million), which he called "an octade" or a "unit of the second class." "Octade octades" (or ten million billions) is called a "unit of the third class," "octade, octade, octades" a "unit of the fourth class," etc.

The writing of large numbers may seem too trivial a

matter to which to devote several pages of a book, but in the time of Archimedes the finding of a way to write big numbers was a great discovery and an important step forward in the science of mathematics.

To calculate the number representing the grains of sand necessary to fill up the entire universe, Archimedes had to know how big the universe was. In his time it was believed that the universe was enclosed by a crystal sphere to which the fixed stars were attached, and his famous contemporary Aristarchus of Samos, who was an astronomer, estimated the distance from the earth to the periphery of that celestial sphere as 10,000,000,000 stadia or about 1,000,000,000 miles.

Comparing the size of that sphere with the size of a grain of sand, Archimedes completed a series of calculations that would give a high-school boy nightmares, and finally arrived at this conclusion:

"It is evident that the number of grains of sand that could be contained in a space as large as that bounded by the stellar sphere as estimated by Aristarchus, is not greater than one thousand myriads of units of the eighth class."[2]

It may be noticed here that Archimedes' estimate of the radius of the universe was rather less than that of modern scientists. The distance of one billion miles reaches only slightly beyond the planet Saturn of our solar system. As we shall see later the universe has now been explored with telescopes to the distance of 5,000,000,000,000,000,000,000 miles, so that the number of sand grains necessary to fill up all the visible universe would be over:

$$10^{100} \text{ (that is, 1 and 100 zeros)}$$

This is of course much larger than the total number of atoms in the universe, $3 \cdot 10^{74}$, as stated at the beginning of this chapter, but we must not forget that the universe is *not* *packed* with atoms; in fact there is on the average only about 1 atom per cubic meter of space.

But it isn't at all necessary to do such drastic things as

2. 10^{63} (i.e., 1 and 63 zeros).

packing the entire universe with sand in order to get really large numbers. In fact they very often pop up in what may seem at first sight a very simple problem, in which you would never expect to find any number larger than a few thousands.

One victim of overwhelming numbers was King Shirham of India, who, according to an old legend, wanted to reward his grand vizier Sissa Ben Dahir for inventing and presenting to him the game of chess. The desires of the clever vizier seemed very modest. "Majesty," he said kneeling in front of the king, "give me a grain of wheat to put on the first square of this chessboard, and two grains to put on the second square, and four grains to put on the third, and eight grains to put on the fourth. And so, oh King, doubling the number for each succeeding square, give me enough grains to cover all 64 squares of the board."

"You do not ask for much, oh my faithful servant," exclaimed the king, silently enjoying the thought that his liberal proposal of a gift to the inventor of the miraculous game would not cost him much of his treasure. "Your wish will certainly be granted." And he ordered a bag of wheat to be brought to the throne.

But when the counting began, with 1 grain for the first square, 2 for the second, 4 for the third and so forth, the bag was emptied before the twentieth square was accounted for. More bags of wheat were brought before the king but the number of grains needed for each succeeding square increased so rapidly that it soon became clear that with all the crop of India the king could not fulfill his promise to Sissa Ben. To do so would have required 18,446,744,-073,709,551,615 grains!

That's not so large a number as the total number of atoms in the universe, but it is pretty big anyway. Assuming that a bushel of wheat contains about 5,000,000 grains, one would need some 4000 billion bushels to satisfy the demand of Sissa Ben. Since the world production of wheat averages about 2,000,000,000 bushels a year, the amount requested by the grand vizier was that of the *world's wheat production for the period of some two thousand years!*

Thus King Shirham found himself deep in debt to his vizier and had either to face the incessant flow of the latter's demands, or to cut his head off. We suspect that he chose the latter alternative.

2. *How To Count Infinities*

In the preceding section we discussed numbers, many of them fairly large ones. But they are still finite and, given enough time, one could write them down to the last decimal.

But there are some really infinite numbers, which are larger than any number we can possibly write no matter how long we work. Thus "the number of all numbers" is clearly infinite, and so is "the number of all geometrical points on a line." Is there anything to be said about such numbers except that they are infinite, or is it possible, for example, to compare two different infinities and to see which one is "larger"?

Is there any sense in asking: "Is the number of all numbers larger or smaller than the number of all points on a line?" Such questions as this, which at first sight seem fantastic, were first considered by the famous mathematician Georg Cantor, who can be truly named the founder of the "arithmetics of infinity."

If we want to speak about larger and smaller infinities we face a problem of comparing the numbers that we can neither name nor write down, and are more or less in the position of a Hottentot inspecting his treasure chest and wanting to know whether he has more glass beads or more copper coins in his possession. But, as you will remember, the Hottentot is unable to count beyond three. Then shall he give up all attempts to compare the number of beads and the number of coins because he cannot count them? Not at all. If he is clever enough he will get his answer by comparing the beads and the coins piece by piece. He will place one bead near one coin, another bead near another coin, and so on, and so on. . . . If he runs out of beads while there are still some coins, he knows that he has more coins than beads; if he runs out of coins with some beads left he knows that he has more beads than coins, and if he comes out even he knows that he has the same number of beads as coins.

Exactly the same method was proposed by Cantor for comparing two infinities: if we can pair the objects of two infinite groups so that each object of one infinite collection pairs with each object of another infinite collection, and no objects in either group are left alone, the two infinities are equal. If however, such arrangement is impossible and in one of the collections some unpaired objects are left, we say that the infinity of objects in this collection is larger, or we can say stronger, than the infinity of objects in the other collection.

This is evidently the most reasonable, and as a matter of fact the only possible, rule that one can use to compare infinite quantities, but we must be prepared for some surprises when we actually begin to apply it. Take for example, the infinity of all even and the infinity of all odd numbers. You feel, of course, intuitively that there are as many even numbers as there are odd, and this is in complete agreement with the above rule, since a one-to-one correspondence of these numbers can be arranged:

$$1 \quad 3 \quad 5 \quad 7 \quad 9 \quad 11 \quad 13 \quad 15 \quad 17 \quad 19 \quad \text{etc.}$$
$$\updownarrow \quad \updownarrow \quad \updownarrow \quad \updownarrow \quad \updownarrow \quad \updownarrow \quad \updownarrow \quad \updownarrow \quad \updownarrow \quad \updownarrow$$
$$2 \quad 4 \quad 6 \quad 8 \quad 10 \quad 12 \quad 14 \quad 16 \quad 18 \quad 20 \quad \text{etc.}$$

There is an even number to correspond with each odd number in this table, and vice versa; hence the infinity of even numbers is equal to the infinity of odd numbers. Seems quite simple and natural indeed!

But wait a moment. Which do you think is larger: the number of all numbers, both even and odd, or the number of even numbers only? Of course you would say the number of all numbers is larger because it contains in itself all even numbers and in addition all odd ones. But that is just your impression, and in order to get the exact answer you must use the above rule for comparing two infinities. And if you use it you will find to your surprise that your impression was wrong. In fact here is the table of one-to-one correspondence of all numbers on one side, and even numbers only on the other:

$$
\begin{array}{ccccccccc}
1 & 2 & 3 & 4 & 5 & 6 & 7 & 8 & \text{etc.} \\
\updownarrow & \updownarrow & \updownarrow & \updownarrow & \updownarrow & \updownarrow & \updownarrow & \updownarrow & \\
2 & 4 & 6 & 8 & 10 & 12 & 14 & 16 & \text{etc.}
\end{array}
$$

According to our rule of comparing infinities we must say that the infinity of even numbers is exactly as large as the infinity of all numbers. This sounds, of course, paradoxical, since even numbers represent only a part of all numbers, but we must remember that we operate here with infinite numbers, and must be prepared to encounter different properties.

In fact in the world of infinity *a part may be equal to the whole!* This is probably best illustrated by an example taken from one of the stories about the famous German mathematician David Hilbert. They say that in his lectures on infinity he put this paradoxical property of infinite numbers in the following words:[3]

Let us imagine a hotel with a finite number of rooms, and assume that all the rooms are occupied. A new guest arrives and asks for a room. "Sorry—says the proprietor—but all the rooms are occupied." Now let us imagine a hotel with an *infinite* number of rooms, and all the rooms are occupied. To this hotel, too, comes a new guest and asks for a room.

"But of course!"—exclaims the proprietor, and he moves the person previously occupying room N1 into room N2, the person from room N2 into room N3, the person from room N3 into room N4, and so on. . . . And the new customer

3. From the unpublished, and even never written, but widely circulating volume: "The Complete Collection of Hilbert Stories" by R. Courant.

receives room N1, which became free as the result of these transpositions.

Let us imagine now a hotel with an infinite number of rooms, all taken up, and an infinite number of new guests who come in and ask for rooms.

"Certainly, gentlemen," says the proprietor, "just wait a minute."

He moves the occupant of N1 into N2, the occupant of N2 into N4, the occupant of N3 into N6, and so on, and so on. . . .

Now all odd-numbered rooms become free and the infinity of new guests can easily be accommodated in them.

Well, it is not easy to imagine the conditions described by Hilbert even in Washington as it was during the war, but this example certainly drives home the point that in operating with infinite numbers we encounter properties rather different from those to which we are accustomed in ordinary arithmetic.

Following Cantor's rule for comparing two infinities, we can also prove now that the number of all ordinary arithmetical fractions like $\frac{3}{7}$ or $7\frac{3}{8}$ is the same as the number of all integers. In fact we can arrange all ordinary fractions in a row according to the following rule: Write first the fractions for which the sum of the numerator and denominator is equal to 2; there is only one such fraction namely: $\frac{1}{1}$. Then write fractions with sums equal to 3: $\frac{2}{1}$ and $\frac{1}{2}$. Then those with sums equal to 4: $\frac{3}{1}$, $\frac{2}{2}$, $\frac{1}{3}$. And so on. In following this procedure we shall get an infinite sequence of fractions, containing every single fraction one can think of. Now write above this sequence, the sequence of integers and you have the one-to-one correspondence between the infinity of fractions and the infinity of integers. Thus their number is the same!

"Well, it is all very nice," you may say, "but doesn't it mean simply that *all* infinities are equal to one another? And if that's the case, what's the use of comparing them anyway?"

No, that is not the case, and one can easily find the infinity

that is larger than the infinity of all integers or all arithmetical fractions.

In fact, if we examine the question asked earlier in this
chapter about the number of points on a line as compared
with the number of all integer numbers, we find that these
two infinities are different; there are many more points on
a line than there are integers or fractional numbers. To
prove this statement let us try to establish one-to-one correspondence between the points on a line, say 1 in. long,
and the sequence of integer numbers.

Each point on the line is characterized by its distance
from one end of the line, and this distance can be written
in the form of an infinite decimal fraction, like 0.73506247-
80056 . . . . or 0.38250375632 . . . .[4] Thus we have to compare the number of all integers with the number of all
possible infinite decimal fractions. What is the difference
now between the infinite decimal fractions, as given above,
and ordinary arithmetical fractions like $\frac{3}{7}$ or $\frac{8}{277}$?

You must remember from your arithmetic that every
ordinary fraction can be converted into an infinite *periodic*
decimal fraction. Thus $\frac{2}{3}$ $=0.66666$ . . . . $=0.(6)$, and $\frac{3}{7}$

$=0.428571 \; \vdots \; 4 \; 28571 \; \vdots \; 4 \; 28571 \; \vdots \; 4 \ldots =0.(428571)$. We

have proved above that the number of all ordinary arithmetical fractions is the same as the number of all integers;
so the number of all periodic decimal fractions must also be
the same as the number of all integers. But the points on a
line are not necessarily represented by periodic decimal
fractions, and in most cases we shall get the infinite fractions in which the decimal figures appear without any
periodicity at all. And it is easy to show that in such case no
linear arrangement is possible.

Suppose that somebody claims to have made such an
arrangement, and that it looks something like this:

4. All these fractions are smaller than unity, since we have assumed the length of the line to be one.

N	
1	0.38602563078 ....
2	0.57350762050 ....
3	0.99356753207 ....
4	0.25763200456 ....
5	0.00005320562 ....
6	0.99035638567 ....
7	0.55522730567 ....
8	0.05277365642 ....
.	....
.	....
.	....
.	....
.	....

Of course, since it is impossible actually to write the infinity of numbers with the infinite number of decimals in each, the above claim means that the author of the table has some general rule (similar to one used by us for arrangement of ordinary fractions) according to which he has constructed the table, and this rule guarantees that every single decimal fraction one can think of will appear sooner or later in the table.

Well, it is not at all difficult to show that any claim of that kind is unsound, since we can always write an infinite decimal fraction that is not contained in this infinite table. How can we do it? Oh, very simply. Just write the fraction with the first decimal different from that of N1 in the table, the second decimal different from that in N2 of the table and so on. The number you will get will look something like this:

	not 3	not 7	not 3	not 6	not 5	not 6	not 3	not 5	etc.
	↓	↓	↓	↓	↓	↓	↓	↓	
0.	5	2	7	4	0	7	1	2	

and this number is not included in the table no matter how far down you look for it. In fact if the author of the table will tell you that this very fraction you have written here stands under the No. 137 (or any other number) in his table you can answer immediately: "No, it isn't the same fraction because the one hundred and thirty seventh decimal in your fraction is different from the other hundred and thirty seventh decimal in the fraction I have in mind."

Thus it is impossible to establish a one-to-one correspondence between the points on a line and the integer numbers, which means that the infinity of points on a line is larger, or stronger, than the infinity of all integer or fractional numbers.

We have been discussing the points on a line "1 in. long," but it is easy to show now that, according to the rules of our "infinity arithmetics," the same is true of a line of any length. In fact, there is the same number of points in lines one inch, one foot, or one mile long. In order to prove it just look at Figure III-1 which compares the number of points on two lines AB and AC of different lengths. To establish the one-to-one correspondence between the points of these two lines we draw through each point on AB a line parallel to BC, and pair the points of intersections as for example D and D^1, E and E^1, F and F^1, etc. Each point on AB has a corresponding point on AC and vice versa; thus according to our rule the two infinities of points are equal.

A still more striking result of the analysis of infinity consists in the statement that: the number of all points on a plane is equal to the number of all points on a line. To prove this let us consider the points on a line AB one inch long, and the points within a square $CDEF$ (Figure III-2).

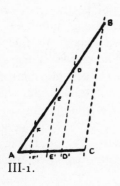

III-1.

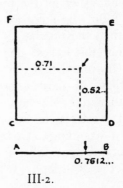

III-2.

Suppose that the position of a certain point on the line is given by some number, say 0.75120386 . . . . We can make from this number two different numbers selecting even and odd decimal signs and putting them together. We get this:

$$0.7108 \ldots$$

and this:

$$0.5236 \ldots$$

Measure the distances given by these numbers in the horizontal and vertical direction in our square, and call the point so obtained the "pair-point" to our original point on the line. In reverse, if we have a point in the square the position of which is described by, let us say, the numbers:

$$0.4835 \ldots$$

and

$$0.9907 \ldots$$

we obtain the position of the corresponding "pair-point" on the line by merging these two numbers:

$$0.49893057 \ldots$$

It is clear that this procedure establishes the one-to-one relationship between two sets of points. Every point on the line will have its pair in the square, every point in the square will have its pair on the line, and no points will be left over. Thus according to the criterion of Cantor, the infinity of all the points within a square is equal to the infinity of all the points on a line.

But the number of all geometrical points, though larger than the number of all integer and fractional numbers, is not the largest one known to mathematicians. In fact it was found that the variety of all possible curves, including those of most unusual shapes, has a larger membership than the collection of all geometrical points, and thus has to be described by the third number of the infinite sequence.

According to Georg Cantor, the creator of the "arithmetics of infinity," infinite numbers are denoted by the Hebrew letter ℵ (aleph) with a little number in the lower right corner that indicates the order of the infinity. The sequence of numbers (including the infinite ones!) now runs:

$$1. \, 2. \, 3. \, 4. \, 5. \ldots \ldots \aleph_1 \, \aleph_2 \, \aleph_3 \ldots \ldots$$

and we say "there are $\aleph_1$ points on a line" or "there are $\aleph_2$ different curves," [5] just as we say that "there are 7 parts of the world" or "52 cards in a pack."

In concluding our talk about infinite numbers we point out that these numbers very quickly outrun any thinkable collection to which they can possibly be applied. We know that $\aleph$ represents the number of all integers, $\aleph_1$ represents the number of all geometrical points, and $\aleph_2$ the number of all curves, but nobody as yet has been able to conceive any definite infinite collection of objects that should be described by $\aleph_3$. It seems that the three first infinite numbers are enough to count anything we can think of, and we find ourselves here in a position exactly opposite to that of our old friend the Hottentot who had many sons but could not count beyond three!

[5] For example from
                0. 735106822548312 . . . . etc.
we make
                0. 71853 . . . .
                0. 30241 . . . .
                0. 56282 . . . .

"Algebra," said E. C. Titchmarsh, Savilian Professor of Geometry at Oxford, "goes to the root of the matter and ignores the casual nature of particular cases." His statement is reminiscent of Bertrand Russell's more celebrated and para-doxical phrase, that mathematics is "the subject in which we never know what we are talking about, nor whether what we are saying is true." The symbolism which is at the core of algebraic method is the basis for both men's statements and accounts for its strength as a mathematical system.

As we have pointed out, two of the earliest and most

fundamental branches of mathematics—geometry and algebra
—developed along different historical lines. Surprisingly,
the Greeks, intellectual masters of the ancient world, who
were so proficient in geometry, made but slight contributions
to its sister science. It was the Hindus, far less enlightened,
who seemed to have grasped the fundamentals of the subject,
and it was probably from them that the Arabs obtained
their initial knowledge of arithmetic and algebra. During the
Middle Ages Jewish scholars from the Moorish universities
in Spain and Gentile merchants in the Levant brought it
into Europe along the trade routes. It was on this foundation of knowledge that the Frenchman, Viete, made the
contributions which Tobias Dantzig calls, "the turning
point in the history of algebra."

Of Number; The Language of Science, by Dantzig, from
which the following selection was taken, no less an authority
than Albert Einstein stated, "This is beyond doubt the
most interesting book on the evolution of mathematics which
has ever fallen into my hands." The author was born in
Russia in 1884, but was educated in the United States and
received his Ph.D. from Indiana University in 1916. He
taught there and at Columbia until he accepted the Professorship of Mathematics at the University of Maryland, from
which he later retired. Dantzig's view of algebra and of
symbols is highly individualistic, particularly in its evaluation
of Principia Mathematica, by Whitehead and Russell. Yet
it is written with a clarity and vigor which is illuminating not
only to trained mathematicians, but also to "cultured
nonmathematicians" for whom his book was written.

SYMBOLS

TOBIAS DANTZIG

ALGEBRA, IN THE BROAD SENSE in which the term is used today, deals with operations upon symbolic forms. In this capacity it not only permeates all of mathematics, but encroaches upon the domain of formal logic and even of metaphysics. Furthermore, when so construed, algebra is as old as man's faculty to deal with general propositions; as old as his ability to discriminate between "some" and "any."

Here, however, we are interested in algebra in a much more restricted sense, that part of general algebra which is very properly called the theory of equations. It is in this narrower sense that the term algebra was used at the outset. The word is of Arabic origin. *Al* is the Arabic article "the," and *gebar* is the verb "to set," to restitute. To this day the word *algebrista* is used in Spain to designate a bone-setter, a sort of chiropractor.

It is generally true that algebra in its development in individual countries passed successively through three stages: the rhetorical, the syncopated, and the symbolic. Rhetorical algebra is characterized by the complete absence of any symbols, except, of course, that the words themselves are being used in their symbolic sense. To this day rhetorical algebra is used in such a statement as "the sum is independent of the order of the terms," which in symbols would be designated by $a+b=b+a$.

Syncopated algebra, of which the Egyptian is a typical example, is a further development of rhetorical. Certain words of frequent use are gradually abbreviated. Eventually these abbreviations become contracted to the point where their origin has been forgotten, so that the symbols have no obvious connection with the operation which they represent. The syncopation has become a symbol.

The history of the symbols + and − may illustrate the point. In mediaeval Europe the latter was long denoted by the full word "minus," then by the first letter "m" duly superscribed. Eventually the letter itself was dropped, leaving the superscript only. The sign "plus" passed through a similar metamorphosis.

The turning-point in the history of algebra was an essay written late in the sixteenth century by a Frenchman, Viète, who wrote under the Latin name Franciscus Vieta. His great achievement appears simple enough to us today. It is summed up in the following passage from this work:

> In this we are aided by an artifice which permits us to distinguish given magnitudes from those which are unknown or sought, and this by means of a symbolism which is permanent in nature and clear to understand— for instance, by denoting the unknown magnitudes by A or any other vowels, while the given magnitudes are designated by B, C, G, or other consonants.

This vowel-consonant notation had a short existence. Within half a century of Vieta's death appeared Descartes's *Géométrie*, in which the first letters of the alphabet were used for given quantities, the last for those unknown. The Cartesian notation not only displaced the Vietan, but has survived to this day.

But while few of Vieta's proposals were carried out in letter, they certainly were adopted in spirit. The systematic use of letters for undetermined but constant magnitudes, the "logistica speciosa" as he called it, which has played such a dominant role in the development of mathematics, was the great achievement of Vieta.

The lay mind may find it difficult to estimate the achievement of Vieta at its true value. Is not the literal notation a mere formality after all, a convenient shorthand at best? There is, no doubt, economy in writing

$$(a+b)^2 = a^2 + 2ab + b^2,$$

but does it really convey more to the mind than the verbal form of the same identity: the square of the sum of two numbers equals the sum of the squares of the numbers, augmented by twice their product?

Again, the literal notation had the fate of all very successful innovations. The universal use of these makes it difficult to conceive of a time when inferior methods were in vogue. Today formulæ in which letters represent general magnitudes are almost as familiar as common script, and our ability to handle symbols is regarded by many almost as a natural endowment of any intelligent man; but it is natural only because it has become a fixed habit of our minds. In the days of Vieta this notation constituted a radical departure from the traditions of ages.

Wherein lies the power of this symbolism?

First of all, the letter liberated algebra from the slavery of the word. And by this I do not mean merely that without the literal notation any general statement would become a mere flow of verbiage, subject to all the ambiguities and misinterpretations of human speech. This is important enough; but what is still more important is that the letter is free from the taboos which have attached to words through centuries of use. The A of Vieta or our present x has an existence independent of the concrete object which it is assumed to represent. The symbol has a meaning which transcends the objects symbolized: that is why it is not a mere formality.

In the second place, the letter is susceptible of operations which enables one to transform literal expressions and thus to paraphrase any statement into a number of equivalent forms. It is this power of transformation that lifts algebra above the level of a convenient shorthand.

Before the introduction of literal notation, it was possible to speak of individual expressions only; each expression, such as $2x+3$; $3x-5$; x^2+4x+7; $3x^2-4x+5$, had an individuality all its own and had to be handled on its own merits. The literal notation made it possible to pass from the individual to the collective, from the "some" to the "any" and the "all." The linear form $ax+b$, the quadratic form ax^2+bx+c, each of these forms is regarded now as a single species. It is this that made possible the general theory of functions, which is the basis of all applied mathematics.

But the most important contribution of the *logistica speciosa,* and the one that concerns us most in this study, is the role it played in the formation of the generalized number concept.

As long as one deals with numerical equations, such as

(I) $x+4=6$ (II) $x+6=4$
$\quad 2x=8$ $\quad\quad 2x=5$
$\quad x^2=9$ $\quad\quad x^2=7,$

one can content himself (as most mediaeval algebraists did) with the statement that the first group of equations is possible, while the second is impossible.

But when one considers the literal equations of the same types:

$$x+b=a$$
$$bx=a$$
$$x^n=a$$

the very indeterminateness of the data compels one to give an *indicated* or *symbolic* solution to the problem:

$$x=a-b$$
$$x=a/b$$
$$x=\sqrt[n]{a}$$

In vain, after this, will one stipulate that the expression $a-b$ has a meaning only if a is greater than b, that $\dfrac{a}{b}$ is meaningless when a is not a multiple of b, and that $\sqrt[n]{a}$ is not a number unless a is a perfect nth power. The very act of writing down the *meaningless* has given it a meaning; and it is not easy to deny the existence of something that has received a name.

Moreover, with the reservation that $a>b$, that a is a multiple of b, that a is a perfect nth power, rules are devised for operating on such symbols as $a-b$; $\dfrac{a}{b}$; $\sqrt[n]{a}$ But sooner or later the very fact that there is nothing on the face of these symbols to indicate whether a legitimate or an illegitimate case is before us, will suggest that there is no contradiction involved in operating on these symbolic beings as if they were bona fide numbers. And from this there is but one step to recognizing these symbolic beings as numbers "in extenso."

What distinguishes modern arithmetic from that of the pre-Vieta period is the changed attitude towards the "impossible." Up to the seventh century the algebraists invested this term with an absolute sense. Committed to natural numbers as the exclusive field for all arithmetic operations, they regarded possibility, or restricted possibility, as an intrinsic property of these operations.

Thus, the direct operations of arithmetic—addition $(a+b)$, multiplication (ab), potentiation (a^b)—were omnipossible; whereas the inverse operations—subtraction $(a-b)$, division $\left(\dfrac{a}{b}\right)$, extraction or roots $\sqrt[b]{a}$, —were possible only under restricted conditions. The pre-Vieta algebraists were satisfied with stating these facts, but were incapable of a closer analysis of the problem.

Today we know that possibility and impossibility have each only a relative meaning; that neither is an intrinsic property of the operation but merely a restriction which human tradition has imposed on the field of the operand. Remove the barrier, extend the field, and the impossible becomes possible.

The direct operations of arithmetic are omnipossible because they are but a succession of iterations, a step-by-step penetration into the sequence of natural numbers, which is assumed *a priori* unlimited. Drop this assumption, restrict the field of the operand to a finite collection (say to the first 1000 numbers), and operations such as $925+125$, or 67×15 become impossible and the corresponding expressions meaningless.

Or, let us assume that the field is restricted to odd numbers only. Multiplication is still omnipossible, for the product of any two odd numbers is odd. However, in such a restricted field addition is an altogether impossible operation, because the sum of any two odd numbers is never an odd number.

Yet, again, if the field were restricted to prime numbers, multiplication would be impossible, for the simple reason that the product of two primes is never a prime; while addition would be possible only in such rare cases as when one of the two terms is 2, the other being the smaller of a couple of twin-primes like $2+11=13$.

Other examples could be adduced, but even these few will suffice to bring out the relative nature of the words possible, impossible, and meaningless. And once this relativity is recognized, it is natural to inquire whether through a proper extension of the restricted field the inverse operations of arithmetic may not be rendered as omnipossible as the direct are.

To accomplish this with respect to subtraction it is sufficient to adjoin to the sequence of natural numbers zero and the negative integers. The field so created is called the general integer field.

Similarly, the adjunction of positive and negative fractions to this integer field will render division omnipossbile.

The numbers thus created—the integers, and the fractions, positive and negative, and the number zero—constitute the rational domain. It supersedes the natural domain of integer arithmetic. The four fundamental operations, which heretofore applied to integers only, are now by analogy extended to these generalized numbers.

All this can be accomplished without a contradiction. And, what is more, with a single reservation which we shall take up presently, the sum, the difference, the product, and the quotient of any two rational numbers are themselves rational numbers. This very important fact is often paraphrased into the statement: the rational domain is *closed* with respect to the fundamental operations of arithmetic.

The single but very important reservation is that of

division by zero. This is equivalent to the solution of the equation $x.o=a$. If a is not zero the equation is impossible, because we were compelled, in defining the number zero, to admit the identity $a.o=o$. There exists therefore no rational number which satisfies the equation $x.o=a$.

On the contrary, the equation $x.o=o$ is satisfied for any rational value of x. Consequently, x is here an indeterminate quantity. Unless the problem that led to such equations provides some further information, we must regard $\dfrac{o}{o}$ as the symbol of *any* rational number, and $\dfrac{a}{o}$ as the symbol of *no* rational number.

Elaborate though these considerations may seem, in symbols they reduce to the following concise statement: if a, b, and c are any rational numbers, and a is not o, then there always exists a rational number x, and only one, which will satisfy the equation

$$ax+b=c$$

This equation is called "linear," and it is the simplest type in a great variety of equations. Next to linear come quadratic, then cubic, quartic, quintic, and generally algebraic equations of any degree, the degree n meaning the highest power of the unknown x in

$$ax^n+bx^{n-1}+cx^{n-2}+ \ldots +px+q=o$$

But even these do not exhaust the infinite variety of equations; exponential, trigonometric, logarithmic, circular, elliptic, etc., constitute a still vaster variety, usually classified under the all-embracing term transcendental.

Is the rational domain adequate to handle this infinite variety? This is emphatically not the case. We must anticipate an extension of the number domain to greater and greater complexity. But this extension is not arbitrary; there is concealed in the very mechanism of the generalizing scheme a guiding and unifying idea.

This idea is sometimes called the principle of permanence. It was first explicitly formulated by the German mathematician, Hermann Hanckel, in 1867, but the germ of the idea was already contained in the writings of Sir William Rowan Hamilton, one of the most original and fruitful minds of the nineteenth century.

I shall formulate this principle as a definition:

A collection of symbols infinite in number shall be called a number field, and each individual element in it a number,

First: If among the elements of the collection we can identify the sequence of natural numbers.

Second: If we can establish criteria of rank which will permit us to tell of any two elements whether they are equal, or if not equal, which is greater; these criteria reducing to the natural criteria when the two elements are natural numbers.

Third: If for any two elements of the collection we can devise a scheme of addition and multiplication which will have the commutative, associative, and distributive properties of the natural operations bearing these names, and which will reduce to these natural operations when the two elements are natural numbers.

These very general considerations leave the question open as to how the principle of permanence operates in special cases. Hamilton pointed the way by a method which he called algebraic pairing. We shall illustrate this on the rational numbers.

If a is a multiple of b, then the symbol $\dfrac{a}{b}$ indicates the operation of division of a by b. Thus $\dfrac{9}{3} = 3$ means that the quotient of the indicated division is 3. Now, given two such indicated operations, is there a way of determining whether the results are equal, greater, or less, without actually performing the operations? Yes; we have the following

$$\text{Criteria} \atop \text{of} \atop \text{Rank} \left\{ \begin{array}{l} \dfrac{a}{b} = \dfrac{c}{d} \quad \text{if} \quad ad = bc \\[2em] \dfrac{a}{b} > \dfrac{c}{d} \quad \text{``} \quad ad > bc \\[2em] \dfrac{a}{b} < \dfrac{c}{d} \quad \text{``} \quad ad < bc \end{array} \right.$$

And we can even go further than that: without performing the indicated operations we can devise rules for manipulating on these indicated quantities:

$$\text{Addition:} \quad \frac{a}{b} + \frac{c}{d} = \frac{ad + bc}{bd}$$

$$\text{Multiplication:} \quad \frac{a}{b} \cdot \frac{c}{d} = \frac{ac}{bd}$$

Now let us *not* stipulate any more that a be a multiple of b. Let us consider $\dfrac{a}{b}$ as the symbol of a new field of mathematical beings. These symbolic beings depend on two integers a and b written in proper order. We shall impose on this collection of *couples* the criteria of rank mentioned above: i.e., we shall claim that, for instance:

$$\frac{20}{15} = \frac{16}{12} \text{ because } 20 \times 12 = 15 \times 16$$

$$\frac{4}{3} > \frac{5}{4} \text{ because } 4 \times 4 > 3 \times 5$$

We shall define the operations on these couples in accordance with the rules which, as we have shown above, are true for the case when a is a multiple of b, and c is a multiple of d; i.e., we shall say for instance:

$$\frac{2}{3} + \frac{4}{5} = \frac{(2 \times 5) + (3 \times 4)}{5 \times 3} = \frac{22}{15}$$

$$\frac{2}{3} \times \frac{4}{5} = \frac{2 \times 4}{3 \times 5} = \frac{8}{15}$$

We have now satisfied all the stipulations of the principle of permanence:

1. The new field contains the natural numbers as a subfield, because we can write any natural number in the form of a couple:

$$\frac{1}{1}, \frac{2}{1}, \frac{3}{1}, \frac{4}{1}, \ldots$$

2. The new field possesses criteria of rank which reduce to the natural criteria when $\frac{a}{b}$ and $\frac{c}{d}$ are natural numbers.

3. The new field has been provided with two operations which have all the properties of addition and multiplication, to which they reduce when $\frac{a}{b}$ and $\frac{c}{d}$ are natural numbers.

And so these new beings satisfy all the stipulations of the principle. They have proved their right to be adjoined to the natural numbers, their right to be invested with the dignity of the name "number." They are therewith admitted, and the field of numbers comprising both old and new is christened the rational domain of numbers.

It would seem at first glance that the principle of permanence leaves such a latitude in the choice of operations as to make the general number it postulates too general to be of much practical value. However, the stipulations that the natural sequence should be a part of the field, and that the fundamental operations should be commutative, associative, and distributive (as the natural operations are), impose restrictions which, as we shall see, only very special fields can meet.

The position of arithmetic, as formulated in the principle of permanence, can be compared to the policy of a state bent on expansion, but desirous to perpetuate the fundamental laws on which it grew strong. These two different objectives—expansion on the one hand, preservation of uniformity on the other—will naturally influence the rules for admission of new states to the Union.

Thus, the first point in the principle of permanence corresponds to the pronouncement that the nucleus state shall set the tone of the Union. Next, the original state being an oligarchy in which every citizen has a rank, it imposes this requirement on the new states. This requirement corresponds to the second point of the principle of permanence.

Finally, it stipulates that the laws of commingling between the citizens of each individual state admitted to the Union shall be of a type which will permit unimpeded relations between citizens of that state and those of the nucleus state.

Of course I do not want the reader to take this analogy literally. It is suggested in the hope that it may invoke mental associations from a more familiar field, so that the principle of permanence may lose its seeming artificiality.

The considerations, which led up to the construction of the rational domain, were the first steps in a historical process called the arithmetization of mathematics. This movement, which began with Weierstrass in the sixties of the last century, had for its object the separation of purely mathematical concepts, such as "number" and "correspondence" and "aggregate," from intuitional ideas, which mathematics had acquired from long association with geometry and mechanics.

These latter, in the opinion of the formalists, are so firmly entrenched in mathematical thought that in spite of the most careful circumspection in the choice of words, the meaning concealed behind these words may influence our reasoning. For the trouble with human words is that they possess content, whereas the purpose of mathematics is to construct pure forms of thought.

But how can we avoid the use of human language? The answer is found in the word "symbol." Only by using a symbolic language not yet usurped by those vague ideas of space, time, continuity which have their origin in intuition and tend to obscure pure reason—only thus may we hope to build mathematics on the solid foundation of logic.

Such is the platform of this school, a school which was founded by the Italian Peano and whose most modern representatives are Bertrand Russell and A. N. Whitehead. In the fundamental work of the latter men, the *Principia Mathematica,* they have endeavored to reconstruct the whole foundation of modern mathematics, starting with clear-cut, fundamental assumptions and proceeding on principles of strict logic. The use of a precise symbolism should leave no room for those ambiguities which are inseparable from human language.

I confess that I am out of sympathy with the extreme formalism of the Peano-Russell school, that I have never acquired the taste for their methods of symbolic logic, that my repeated efforts to master their involved symbolism have invariably resulted in helpless confusion and despair. This personal ineptitude has undoubtedly colored my opinion—a powerful reason why I should not air my prejudices here.

Yet I am certain that these prejudices have not caused me to underestimate the role of mathematical symbolism. To me, the tremendous importance of this symbolism lies not in these sterile attempts to banish intuition from the realm of human thought, but in its unlimited power to aid intuition in creating new forms of thought.

To recognize this, it is not necessary to master the intricate technical symbolism of modern mathematics. It is sufficient to contemplate the more simple, yet much more subtle, symbolism of language. For, in so far as our language is capable of precise statements, it is but a system of symbols, a rhetorical algebra *par excellence.* Nouns and phrases are but symbols of classes of objects, verbs symbolize relations, and sentences are but propositions connecting these classes. Yet, while the word is the abstract symbol of a class, it has also the capacity to evoke an image, a concrete picture of some representative element of the class. It is in this dual function of our language that we should seek the germs of the conflict which later arises between logic and intuition.

And what is true of words generally is particularly true of those words which represent natural numbers. Because they have the power to evoke in our mind images of concrete collections, they appear to us so rooted in firm reality as to be endowed with an absolute nature. Yet in the sense in which they are used in arithmetic, they are but a set of abstract symbols subject to a system of operational rules.

Once we recognize this symbolic nature of the natural number, it loses its absolute character. Its intrinsic kinship with the wider domain of which it is the nucleus becomes evident. At the same time the successive extensions of the number concept become steps in an inevitable process of natural evolution, instead of the artificial and arbitrary legerdemain which they seem at first.

One of the oldest branches of mathematics continues to be among the most vigorous and important. A thousand years before Christ, the Egyptians had learned to do relatively difficult problems in geometry. The Babylonians too had studied its principals. As we have seen, it remained for the Greeks to bring it to its full flowering in ancient times. This majestic body of knowledge, built step by step for centuries, was passed on to succeeding generations by the Elements.

But modern geometry is far removed from Euclid's thirteen books. They were based on reason as it was known to the philosophers of the period, on "common sense" and "self-evident" axioms and principles of logic. Modern geometry has little to do with the ancient conception of common sense and, certainly to the unsophisticated, is anything but self-evident. G. N. Lewis, the noted physical chemist, lures us with deceptive simplicity into a consideration of the troublesome *parallel postulate whose abandonment resulted in the geometries of Lobachevski, Bolyai, and Riemann. As we read, we may on occasion sympathize with the canoeist's wife in the allegory Lewis relates. But the developments he describes are among the most important in the history of mathematics. On them rest such diverse studies as the science of navigation and the rules of transformation.*

GEOMETRIES

GILBERT N. LEWIS

ONCE UPON A TIME the South Pacific Ocean was so overcast by clouds for a number of generations that the islanders had only a vague tradition of sun and stars. Nevertheless, their civilization grew. Their coconut plantations became so valuable that they made accurate land measurements with lines of coconut fiber, and these methods of measurement were codified by a great geometer, named Uli, so that their children were taught the same kind of theorems and corollaries as our children are now taught. They ventured out but short distances in their canoes until they discovered the lodestone and the compass. Then their navigators became bolder and finally they had mapped out a great part of the Southern Pacific. One of their maps is shown in Figure III-3. A vertical line represents the path of a canoe going directly with the compass, or as we say, north and south, and the horizontal line the path of a canoe traveling in a transverse direction, neither to the north nor to the south. The lines were placed, as they supposed, at the same distance apart, namely, ten days' paddling. They had therefore what we call a Mercator's projection. But they had no idea of a spherical earth; to them it was quite flat.

The base line of their map ran between the two islands

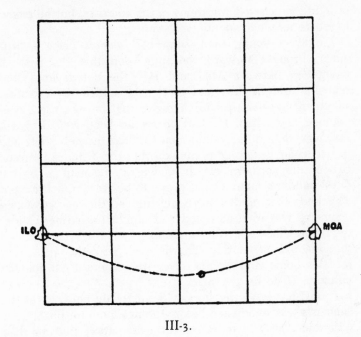

III-3.

called Ilo and Moa, forty days' paddling apart. Once a
number of canoes set out on this voyage leaving Moa at the
same time, each paddling by compass. But one was soon
lost from the others and came upon a rocky island which
the captain of the canoe knew to be far south of the
regular route. He discovered that an iron amulet he was
wearing had disturbed his compass, and having discarded
this he set course by his compass for Ilo. Much to his
surprise, he arrived there before any of the other canoes.
Being of an experimental turn of mind he returned by the
same route, and once more beat all his competitors. The
story spread, others made similar trials, and finally they
found that the quickest path between the islands was the
one shown on the map by the dotted line. It was at first
supposed that ocean currents made the circuitous path
the quicker one, but no such currents could be discerned.
Then it was proposed that some great submerged mountain

of lodestone caused variations in the compass, but all efforts to locate its whereabouts were vain.

Finally a young man, who had always been a little queer, brought forward the suggestion that the path of navigation between Moa and Ilo (the dotted line) was really a straight line, and to this he added several other statements so heretical in character that he was tried, condemned and eaten. He had no sooner been well digested, however, than other young men took up the new idea, and in particular tested his prediction that the distance between two of the northern islands shown on the map to be ten days' paddling apart would prove to be really much longer. This and other predictions being fulfilled the new advanced geometry was soon accepted and taught to mariners under the name Uliao, which means "more than Uli." It was found that the old elementary Uli was still perfectly satisfactory for the home and thence arose the saying "Uli for the minnow, Uliao for the shark."

I must not leave my story without telling how one of the canoeists was summoned before the headman for beating his wife too noisily. He stated in extenuation that he had sat up for two hours explaining to her the principle of direct paddling between Moa and Ilo, and at the end she had said, "How very interesting, but why is it not quicker to go straight across?"

I shall not dwell upon the moral of this allegory, but I shall ask you to think of it now and then, as it seems to me to contain a good deal of the philosophy of science that I am attempting to present to you. In particular, I have tried to bring out the idea that the geometry of navigation is a two-dimensional, non-Euclidean geometry, and would very likely have been regarded as the geometry of the *plane* if it had been familiar before the earth was known to be spherical.

The geometry of Euclid has come down to us as one of the great monuments of Greek thought. For over two thousand years it has been employed almost without change as a textbook of elementary instruction. For centuries it was deemed a heresy to dispute its validity. Yet during that period it has shown its vitality by growth and change; otherwise it would have come to us only like a fossil of ancient thought, or like some brilliant butterfly impaled in a museum case.

If we now claim a better understanding of geometry than was possessed by the Greeks, it is not through any lack of admiration of their stupendous achievement; nor is it that we believe that we are better thinkers, or that our ideas have yet reached anything like completion. I suspect that it is only the very young vine that says, "When I grow big I shall reach the sun." The older thinks only, "I climb a little higher day by day."

The geometry of Euclid is an excellent example of a structure based upon several clear-cut and distant logical levels. In his "common notions," such as that the whole is greater than any one of its parts, Euclid presents a number of general rules of thought (or rules of language) which are useful in many other fields besides the limited field of geometry. In his postulates he states the particular rules of his geometry; and finally in his definitions he displays the subject matter,—points, lines, triangles, circles and the like. Nowadays it is the fashion to leave these elements undefined, but that is a matter of taste. No word can be completely defined, but it is sometimes useful to illustrate the meaning we desire to convey by means of examples or analogies or pictures.

It is, however, of the postulates that I wish to speak. Let me remind you of their content.

1] A straight line may be drawn from any point to any point.

2] A straight line may be produced continuously in a straight line.

3] A circle may be described with any center or distance.

4] All right angles are equal.

5] If a straight line falling on two straight lines makes the interior angles on the same side less than two right angles, the two straight lines, if produced indefinitely, meet on that side on which are the angles less than the two right angles.

If after neatly packing for a journey we should discover some necessary object and add it as an ungainly package to our carefully packed trunks and boxes, it would give the same impression that Euclid produces when after the rest of his simple and concise utterances he introduces the awkward sentence that appears in some editions as the last postulate, in others as the last of the common notions, namely, "If a straight line falling on two straight lines, etc."

This postulate which seems to have been and probably was inserted without much polish, at the last moment, as though it had been hoped to dispense with it, is nevertheless one of the greatest of Euclid's creations, for without such a postulate Euclidean geometry could not exist. There is nothing more interesting in the history of science than the record of the repeated attacks made on this postulate in the hope of reducing it to a proposition deducible from the other four, and which finally resulted in the discoveries by Lobachevski,[1] Bolyai and Riemann of other geometries, all of which are deducible from the material used by Euclid, with the single elimination of this one postulate.

For our purpose we may express the postulates of Euclid in simpler form.

1] Through two points there is a unique line and this may be called the straight line.

2] About any point circles may be described.

3] Through any point outside a line one and only one parallel can be drawn.

It is the last postulate which is abandoned in the non-Euclidean geometries of which I have spoken. In the geometry of Lobachevski more than one parallel can be drawn; in that of Riemann none can be drawn. The geometry of

1. Those who are interested in the problem of the objective and subjective will find it worthy of note that the two geometries which were published independently and almost simultaneously, one by the Russian Lobachevski, and the other by the Hungarian Bolyai, were so nearly alike that they seem like different drafts of the same composition. Similarly Hamilton and Grassmann wrote at the same time those papers which were to become the foundation of modern vector analysis. We cannot avoid the thought that having embarked upon a certain line of mathematical inquiry, while we appear to have preserved the utmost of personal freedom, we seem bound to follow certain paths and to make and remake certain discoveries, just as we do in physics or chemistry.

navigation which we discussed in our allegory belongs to the last-named class. In it the straight line is what we call a great circle, and, as you know, no two great circles are parallel to each other.

The geometry of navigation would be different on a small planet and on a large one. It would depend upon what we call the curvature of the surface. And all of these non-Euclidean geometries of which I have spoken involve a certain absolute magnitude, which by analogy is called curvature. So also there are called "curved" geometries as distinguished from the "flat" geometry of Euclid, although it is by no means implied that they find applicability only upon curved surfaces. As that absolute magnitude to which I referred becomes smaller and smaller, the geometries approach the geometry of Euclid as a limit. Moreover, the geometry of any given region approaches more nearly to the Euclidean as the region considered becomes smaller, just as in our allegory the South Sea islanders found their elementary geometry sufficient for the home.

Are these geometries true and is Euclid's geometry false? This is a question which no longer conveys any meaning to our minds. Is chess true? Provided that a geometry contains within itself no inconsistencies or absurdities, then we regard it as true just in so far as it is interesting or useful. Certainly the laws of navigation are true and the only two-dimensional geometry that they fit is a non-Euclidean geometry. This is one of the so-called geometries of positive curvature, and all of the geometries of this class have attained enormous importance owing to recent theories of gravitation. The geometry of so-called negative curvature, typified by the original non-Euclidean geometry of Loba-chevski and Bolyai, has as yet found application only in minor ways, but its simplicity and elegance make us almost certain that it also will eventually be of great utility.

These curved geometries are already familiar to many of you, and I do not propose to discuss them further, but rather to call your attention to two geometries which have come into very recent notice, of which one bids fair to rival in importance Euclidean geometry itself. The parallel postulate of Euclid has been the storm center of geometry for centuries, but little attention has been paid to his postulate regarding circles. It is, however, by retaining the parallel postulate and abandoning the circular postulate that the two new geometries of which I am about to speak are obtained.

First, however, let me show how in one respect modern geometrical thought has run ahead of Euclid's. The chief problem of geometry is to compare the metrical properties of one figure with those of another. Language has its words and the rules for using words that we call grammar. Chess has its pieces and its moves, arithmetic its numbers and its operations upon numbers; geometry has its figures and its methods of comparing figures. As the rules of chess apply, not to all games, but to a single game, so the rules of Euclid apply, not to all geometries, but only to one geometry.

Now the method which Euclid used for comparing one figure with another was only a slight idealization of the method of cutting the figures out of paper and moving them about to see how nearly they fit one another. Any such method of transposing a figure may evidently be divided into two movements, one of sliding without turning, and one of turning without sliding, and we shall see that these two types of movement are closely connected respectively with the parallel postulate and the circular postulate of Euclid.

We now proceed more circumspectly than Euclid did, and we have gone further in idealizing the physical process of moving a figure cut out of wood or paper. We do not wish to be limited by the particular properties or by the imperfections of such substances, nor do we wish to be influenced by our intuition. Instead we set up a body of rules according to which we agree to be governed as long as we are playing a particular game of geometry. Thus we make rules for what we call a *transformation* whereby a figure is reproduced in (rather than moved to) another part of our diagram, and we ordinarily make the rules such that the figure thus produced has the same intrinsic metrical properties as its original. Thus the area, the length of a certain side, the angle between two sides, we shall agree to call the same in the new figure as in the old. It will suffice to consider two kinds of transformation, one of which we may call parallel shift, and the other, rotation.

The first of these is so defined that every line produced by the parallel shift is parallel to its original. This is illustrated in Figure III-4 where ABDC is the original figure, and A'B'D'C' is its reproduction. Such a transformation may be made in a single step, as shown in the figure, or in a succession or "group" of steps, each of which is itself a parallel shift.

Since in any parallel shift a line goes over into another line of equal length, it is possible not only to compare the length of any part of the line AB with the length of any other part of it, but also with the length of any part of A'B'. In other words, we may compare a length along any one line with a length along any other parallel line, but our rules so far give us no information whatever as to the relative

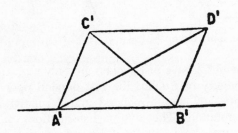

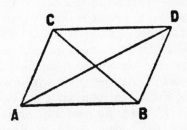

III-4.   *The Parallel Shift.*

length of AB and AD. Certainly AD looks longer, but our eye is trained only in Euclidean methods of measurement, and we must now agree to form no such judgment until we have set up definite rules for the comparison of nonparallel lines.

In the meantime, if we had the time, we might develop the full geometry of the parallel postulate and the associated parallel transformation, and thus obtain a large number of theorems regarding triangles and parallelograms which are to be found in the geometry of Euclid, but which are also contained in the other flat geometries which I am going to describe.

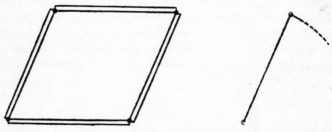

III-5.   *Parallel Ruler and Compass.*

The theorems of which I have just spoken are only a part, however, of the geometry of Euclid; the rest depend upon the circular postulate and the associated transformation, which is called a rotation. The two transformations of Euclidean geometry may be carried out by means of the two instruments shown in Figure III-5. The one on the left is the parallel ruler made of four rods hinged together; the one on the right is the compass, which in its simplest form is a piece of string which is kept taut while one end is fixed. The first instrument we may use in all our flat geometries, but the second must be modified in accordance with the rules of transformation which we decide to employ in place of Euclidean rotation.

Let me show you now an instrument which I may call a non-Euclidean compass (Figure III-6). It consists of two

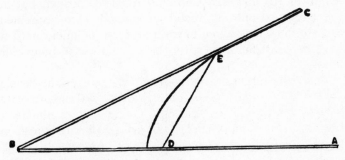

III-6.   *Non-Euclidean Compass.*

rods, AB and BC, of which the first is stationary and the
second is allowed to move about a hinge at B. A cord is
attached to the two rods at the points C and D, and now
if a piece of chalk is pressed against the string so as to keep
the upper part of the string coincident with the rod BC and,
maintaining this condition, is moved slowly downward, we
trace the heavy curve in the figure. If I had used a more
complicated but also more complete apparatus, my compass
would have drawn simultaneously the curves of Figure
III-7, which do not look like circles, but look to the Euclidean eye like hyperbolas, with dotted lines as *asymptotes*
(lines which are approached indefinitely, but never reached).
And yet we may define the rules of a transformation which
will make the line OA go into OB and OB into OC, while
the line OA′ goes into OB′, which in turn goes to OC′; and
I shall venture to call this a rotation.

Here I follow the distinguished precedent of Humpty
Dumpty in *Through the Looking-Glass* who said, "When
I use a word it means just what I choose it to mean—
neither more nor less." Of course Humpty Dumpty and I
are both wrong, for it is impossible to free a word from all
its countless implications, unless we make it out of the
whole cloth, as the word "gas" was once supposed to have
been made. But here the point is that it seems wiser to
emphasize the resemblances rather than the differences
between the old and new types of rotation. You may prefer
to call the new process absurd or irrational rotation, and in-

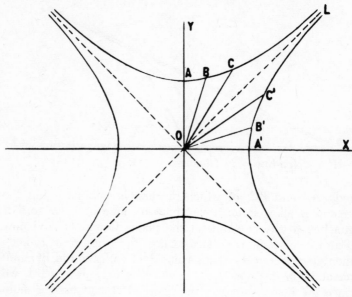

III-7. *Asymptotic Rotation.*

deed extending the word "rotation" to cover this new asymptotic rotation is very much like extending the word "number," which referred originally to integers, to include fractions, and later irrational fractions.

Enlarging the meaning of rotation implies also an extension of the meaning of distance, because we have agreed to abide by the rules of our transformation in comparing the length of two nonparallel lines. Since in our transformation OA goes into OB and OB into OC, we must say that A, B and C are equidistant from O. Thus we lay down the rules of a geometry which is as simple and beautiful as the geometry of Euclid, and which has such important applications that we may before long see this geometry taught by the side of Euclid in our schools.

Finally, we may set up rules for a third flat geometry in which the curve of rotation is neither circular nor asymptotic, but merely a straight line. The three types of curves of rotation are shown in Figure III-8. The new kind of

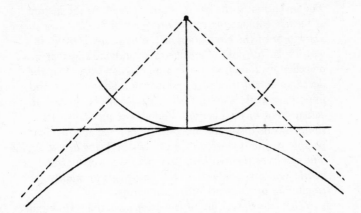

III-8. *The Three Types of Rotation.*

rotation corresponds to what physicists call a shear; it is like the movement of a pack of cards when the top slides and the bottom remains stationary. The geometry of shear rotation defined by such a transformation lacks the fullness and complexity of either the Euclidean or the asymptotic varieties. It might for this reason be called a "degenerate" geometry.

Thus by taking liberties successively with two of Euclid's postulates it is possible to obtain the four additional types of geometry shown at either side of the central Euclidean geometry in the following tabulation:

No curvature	No curvature	No curvature	Negative curvature	Positive curvature
Asymptotic rotation	Shear rotation	Circular rotation	Circular rotation	Circular rotation

The two on the right are the older non-Euclidean geometries, those on the left the newer ones. There may be other branches of mathematics which deserve to be called geometries, but these five, together with their hybrids, constitute the great family of Euclididæ.

The invention of the calculus was an event in the history
of science comparable to the discovery of evolution by
Darwin or of the circulation of the blood by Harvey. It
was similar to those other scientific landmarks in that it was
preceded by a long, slow accretion of knowledge; that the
developments it suggested resulted in a flood of new and
important work; and that the idea for it was in the air—the
actual event of discovery seemed almost inevitable. Its
simultaneous discovery by Newton and Leibnitz is comparable to the discovery of evolution by Darwin and Wallace.
The claims of the two biologists, however, were pursued with
gentle courtesy, while those for mathematical priority
resulted in bitter chauvinistic contention.

What was the nature of the great invention? How did
it aid the development not only of mathematics but of all
the other sciences? What was the basic idea of the "variable"
which lies at the root of the calculus? How did its concepts
of change, of flow, of movement, illuminate all of nature?
These are the questions which are discussed below.

Edward Kasner was a distinguished geometer. He was
born at New York in 1878 and educated there and in Germany. After a long period of teaching at Columbia, he was
appointed Adrian Professor of Mathematics in 1937. A rare
and gifted teacher, he died in New York in 1955. A short
note on his collaborator, James Newman, precedes
Newman's article on Albert Einstein, appearing on page 78.

CHANGE AND CHANGEABILITY—THE CALCULUS

EDWARD KASNER AND JAMES NEWMAN

> *The ever-whirling wheele*
> *Of* Change, *the which all mortall things doth sway.*
> —SPENSER

> *People used to think that when a thing changes,*
> *it must be in a state of change, and that when a*
> *thing moves, it is in a state of motion. This is*
> *now known to be a mistake.*
> —BERTRAND RUSSELL

"EVERYONE WHO UNDERSTANDS the subject will agree that even the basis on which the scientific explanation of nature rests is intelligible only to those who have learned at least the elements of the differential and integral calculus. . . ." These words of Felix Klein, the distinguished German mathematician, echo the conviction of everyone who has studied the physical sciences. It is impossible to appraise and interpret the interdependence of physical quantities in terms of algebra and geometry alone; it is impossible to proceed beyond the simplest observed phenomena merely with the aid of these mathematical tools. In the construction of physical theories, the calculus is more than the cement which binds the diverse elements of the structure together, it is the implement used by the builder in every phase of the construction.

Why is this branch of mathematics peculiarly suited for the precise formulation of natural phenomena? What powers can be attributed to the calculus that are not also shared by geometry and algebra?

Our most common impression of the world, whether erroneous or not, is its ever-changing aspect. Nature, as well as the artifacts we have invented to master it, seems to be in constant flux. Even the "absolutes"—space and time— contract and expand incessantly. Night and day repeatedly flow into one another, setting forth the vicissitudes of the seasons. Everywhere there is motion, flow, cycles of birth, death and regeneration. Everywhere the pattern moves.

The word "calculus" originally meant a small stone or pebble; it has acquired a new connotation. The calculus may be regarded as that branch of mathematical inquiry which treats of *change* and *rate of change*. The comfort with which one rides in an automobile is made possible, in part at least, by the calculus. While the planets would continue in their paths without the calculus, Newton needed it to prove that their orbits about the sun are ellipses. Shrinking from the celestial to the atomic, the solution of the very same equation used by Newton to describe the motion of the planets determines the trajectory of an alpha particle which bombards an atomic nucleus. By means of the formula which relates the distance traversed by a moving body to the time elapsed, the velocity of the body, as well as its acceleration, at every instant of time is determined by the calculus.

Each of the above illustrations, whether simple or complex, involves change and rate of change. Without their exact mathematical enunciation none of the problems described would have meaning, much less be solvable. Thus, a mathematical theory has been created which takes cognizance of immanent and ubiquitous changes of pattern and which undertakes to examine and explain them. That theory is the calculus.

The calculus does not differ from other mathematical theories; it did not spring full-grown from the genius of any one man. Rather was it developed from a consideration of numerous questions essayed and successfully answered by the predecessors of Newton and Leibnitz.

The advent of analytic geometry furnished a powerful stimulus to the invention of the calculus, for the pictorial representation of a function revealed many interesting features. Kepler had noticed that, as a variable quantity approaches its maximum value, its rate of change becomes

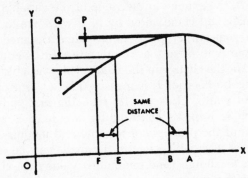

III-9. *The rate of change of a variable quantity is smaller at a maximum point than elsewhere.*

less than at any other value. It continues to choke off until, at the maximum value of the variable, the rate of change is zero.

In the above diagram, the values assumed by a variable quantity are measured by the distance from the straight line (the x axis) to the curve. The maximum value of the variable quantity (the greatest distance from the x axis to the curve) is attained at the point labeled A; when moving slightly either to the right or to the left of A, for instance to

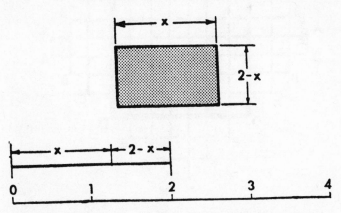

III-10. *Using the scale, the perimeter of the rectangle is clearly 4 units.*

the point B, the change in the value of the variable quantity is very small, and is measured by P. If we move to the right or left of some other point E the same distance that we moved from A to B, so that the distance EF is equal to the distance AB, the change in the value of the variable quantity in the neighborhood of E is measured by Q. But obviously, this second width, Q, is greater than the first width, P. In this, which is Kepler's contribution, we have a geometric illustration of the principle of maxima and minima: the rate of change of a variable quantity is smaller in the neighborhood of its maximum (and minimum) value than elsewhere. In fact, at the maximum and minimum values, the rate is zero.

Pierre de Fermat, who shares with Descartes the distinction of discovering analytic geometry, was one of the first mathematicians to devise a general method applicable to the

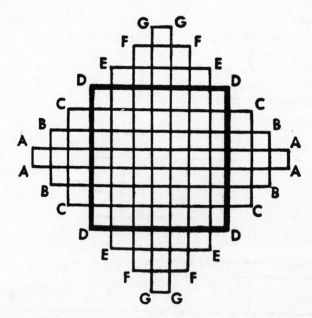

III-11. *The perimeter of each of the seven rectangles viz. AAAA, BBBB, CCCC, etc., is the same. But obviously the rectangle of maximum area is the square DDDD.*

solution of problems involving maxima and minima. His method, used as early as 1629, is substantially that applied today to problems of this type. Let it be required to draw a rectangle such that the sum of the sides is four inches and such that the area * shall be a maximum. If we denote one side of the maximum rectangle by x, the adjacent side, as may be seen from Fig. III-11, will be $2 - x$; and the area of the rectangle will be $x(2 - x)$. If the side x is increased by a small amount E, the side $2 - x$ will have to be diminished by E in order to maintain a constant perimeter. The new area will then be $(x + E)(2 - x - E)$. Since the original area was a maximum, this slight alteration in the relation of the sides can have produced only a slight change in the area. Thus, considering the two areas *approximately* equal, we have

$$x (2 - x) = (x + E) (2 - x - E)$$

whence $\qquad 2x - x^2 = 2x - x^2 - Ex + 2E - Ex - E^2.$

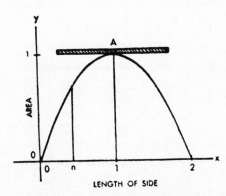

III-12. The curve is a parabola representing the areas of all rectangles whose perimeter is 4 units long.

Erect a perpendicular at any point n along the x axis to the curve. The length of this perpendicular will be the area of the rectangle, one side of which equals n. The maximum area corresponds to the point A on the graph, i.e., the perpendicular erected at x = 1. Thus the rectangle of maximum area, with a perimeter = 4, has a side = 1 and is therefore a square.

*The area of a rectangle is the product of two adjacent sides.

Subtracting $2x - x^2$ from both sides of this equation and factoring:

$$0 = 2E - 2Ex - E^2$$
$$0 = E(2 - 2x - E).$$

But E is *not* equal to zero, therefore the other factor $(2 - 2x - E)$ must be zero:

$$0 = 2 - 2x - E.$$

As smaller and smaller values are taken for E, (i.e., as the altered rectangle approaches closer and closer to the original maximum rectangle) the expression on the right-hand side of the equation approaches closer and closer to the expression obtained by setting E equal to zero, namely, $2 - 2x$. Solving this resulting equation:

$$0 = 2 - 2x$$

we find that: $x = 1$; or, in terms of the original problem, the rectangle with the maximum area is a square.

It is well to note that the area of the rectangle is a function of the lengths of the sides, and this function can be portrayed by a curve (Figure III-12).

The highest point of this curve is at $x = 1$. This is the *maximum* of the function. To use a crude analogy, since this point is neither "uphill" nor "downhill," a small steel ball would be in equilibrium, or a ruler could be balanced at this point. If we think of a straight line being "balanced" at this point, such a line would be called the "tangent to the curve." The interesting fact is that the tangent to a curve at its maxima and minima points will always be horizontal (Figure III-13). To this idea, so important in the calculus, we shall return later.

A tiny flame, lit by Archimedes and his predecessors, burst forth into new brilliance in the intellectually hospitable climate of the seventeenth century to cast its light over the entire future of science. The fertile concept of *limit* revealed its full powers for the first time in the development of the differential calculus.

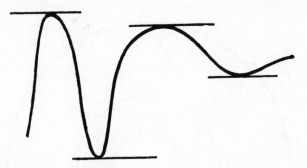

III-13. *The horizontal lines are tangent to the relative maxima and minima of the curve.*

We are acquainted with the limit of a variable quantity. The sequence of numbers 0.9, 0.99, 0.999, 0.9999 . . . converges to the limiting value 1. The series $1 + \frac{1}{2} + \frac{1}{4} + \frac{1}{8} + \frac{1}{16} + \ldots$ converges to the limiting value 2. Nor are geometric examples unfamiliar. If a regular polygon is inscribed in a circle, the difference between the perimeter of the polygon and the circumference of the circle can be made as small as one wishes merely by taking a polygon with a sufficient number of sides. The limiting figure is the circle, the limiting area, the area of the circle.

In these instances, there is no difficulty in determining the limit; this is the exception, however, not the rule. Usually, a formidable mathematical procedure is required to determine the limit of a variable quantity. Consider this: Draw a circle with a radius equal to one. In it inscribe an equilateral triangle. In the triangle inscribe another circle; in the second circle, a square. Continue with a circle in this square, and follow with a regular five-sided figure in the circle. Repeat this procedure, each time increasing the number of sides of the regular polygon by one.

At first glance, one might suppose that the radii of the shrinking circles approach zero as their limiting value. But this is not so; the radii converge to a definite limiting value different from zero. As an explanatory clue, it need only be remembered that the shrinking process itself ap-

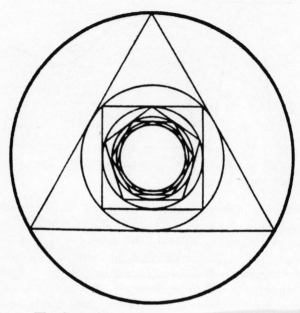

III-14. *The diminishing radii approach a limit approximately* $\frac{1}{12}$ *that of the radius of the first circle.*

proaches a limit as the circles and inscribed polygons become approximately equal. The limiting value of the radii is given by the infinite product:

$$\text{Radius} = \cos\frac{\pi}{3} \times \cos\frac{\pi}{4} \times \cos\frac{\pi}{5} \times \ldots \times \cos\frac{\pi}{(n+2)}$$

Closely related to this problem is the one of circumscribing the regular polygons and the circles instead of inscribing them.

Here it would seem that the radii should grow beyond bound, become infinite. This, too, is deceptive, for the radii of the resulting circles approach a limiting value given by the infinite product:

III-15. *The increasing radii approach a limit approximately 12 times that of the original circle.*

$$\text{Radius} = \cfrac{1}{\cos\dfrac{\pi}{3} \times \cos\dfrac{\pi}{4} \times \cos\dfrac{\pi}{5} \times \ldots \times \cos\dfrac{\pi}{(n+2)}}$$

Interestingly enough, the two limiting radii are so related that one is the reciprocal of the other.

So much for the limit of a variable quantity. Let us now turn to the limit of a function, recalling briefly the meaning of function.* It is often found that two variable quantities are so related that to each value of one there corresponds

---

*Though we have done this before, the notion of function is so important, so all-pervading in mathematics, that it is worth going over again.

a value of the other. Under this condition, the two variable quantities are said to be functions of one another, or functionally related. Thus, the force of attraction (or repulsion) between two magnets is a function of the distance between them. The greater the distance between the magnets, the less the force; the less the distance, the greater the force.

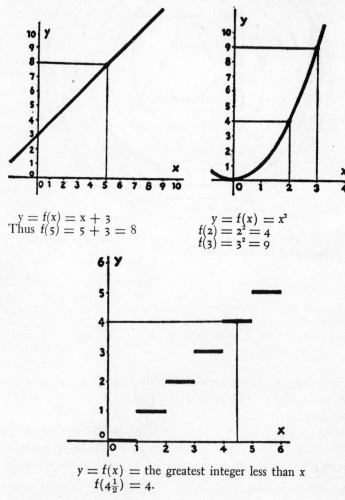

$$y = f(x) = x + 3$$
Thus $f(5) = 5 + 3 = 8$

$$y = f(x) = x^2$$
$$f(2) = 2^2 = 4$$
$$f(3) = 3^2 = 9$$

$$y = f(x) = \text{the greatest integer less than } x$$
$$f(4\tfrac{1}{2}) = 4.$$

III-16.   *Portraits of three different functions.*

If the distance is permitted to assume arbitrary values, it may be considered as an independent variable. The force then becomes the dependent variable, dependent upon the distance (and the functional relation) and is uniquely determined by assigning values to the independent variable. In functional relations, the letter x usually denotes the independent variable, the letter y the dependent variable. The dependency "y is a function of x" is written symbolically:

$$y = f(x).$$

The equation $y = f(x)$ determines a value of y for every value of x. Each pair of values which satisfies this equation is considered as the Cartesian coordinates of a point in a plane; the curve depicting the function is composed of all such points.

In discussing the concept "limit of a function," let us study specifically the function $y = \dfrac{1}{x}$, represented graphically in Figure III-17.

The value of the function at the point $x = \frac{1}{2}$ is $y = f(\frac{1}{2}) = 2$. This value is graphically represented by the distance from the point on the x axis, $\frac{1}{2}$ unit to the right of the origin, to the curve. Likewise, the value of the function at each point along the curve is represented by its distance from the x axis.

For the function $y = \dfrac{1}{x}$, take two neighboring points, $x = \frac{1}{4}$, and $x = \frac{1}{2}$. As the independent variable moves along the x axis from the point $x = \frac{1}{4}$ to $x = \frac{1}{2}$, the dependent variable is "forced" along the curve from the point $y = f(\frac{1}{4}) = 4$ to $y = f(\frac{1}{2}) = 2$. In other words, as the independent variable x approaches as its limit the value $\frac{1}{2}$, the dependent variable, the function, approaches as its limit the value 2. Generally, as an independent variable x

approaches a value *A*, its dependent variable *y* (the function of *x*) approaches a value *B*. Thus, the limit of *f(x)*, as *x* approaches *A*, is *B*. This is what is meant by "limit of a function."

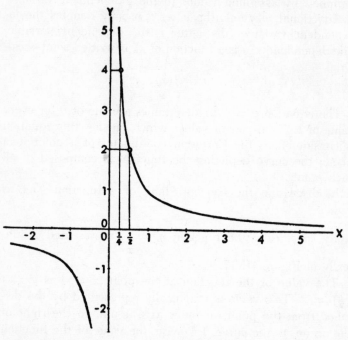

III-17. *Graph of the function* $y = \dfrac{1}{x}$

Recalling the example of the steel rod flexed under a weight, we may construct a parallel dictionary of terms.

MATHEMATICS	PHYSICS
Independent variable, *x*.	Amount of weight.
Dependent variable, *y*.	Amount of bend of steel rod.
Function is the relation be-tween *x* and *y*.	Function is the relation be-tween the weight and the degree of bend.

Increase or decrease of x (i.e., change).	Addition or diminution of weight (i.e., change).
Increase or decrease of y (i.e., change).	Increase or decrease in the degree of bend of the steel rod (i.e., change).
Limiting value of y (the function of x) equals a number.	Limiting value of degree of bend (function of the weight) equals a position.

With the concepts limit, function, and limit of a function in mind, there remains to define an idea embracing all three—"rate of change."

Consider the determination of the speed of a moving body at a given instant of time. A bomb is dropped from a stationary airship at an altitude of 400 feet. Five seconds will elapse before it hits the ground. Its average speed is

thus $\dfrac{400 \text{ feet}}{5 \text{ seconds}} = 80$ feet per second. Hence, the average

rate of change of distance with respect to time is 80. We are aware, however, from the most elementary knowledge of physics that a body gathers speed as it falls. Throughout the fall the bomb was not moving at a constant rate of 80 feet per second; the speed with which it fell varied from point to point, increasing at each successive instant (disregarding air resistance). Suppose, for the sake of simplicity, we limit our interest to the speed of the bomb at the exact moment of striking the ground. Evidently, its speed during the last second before striking will give a fair approximation to its speed at the instant of striking. The distance covered during this last second being 144 feet, the rate of change of distance with respect to time is 144. If we now take smaller and smaller intervals of time, we may expect to obtain closer and closer approximations to the speed of the projectile at the moment of impact. In the last half second, the distance covered was 76 feet, so that the speed was 152 feet per second. The table lists the intervals of time, the distance covered in those intervals, and the average speed over each interval. It is readily seen that as the interval of time approaches zero, we obtain the approximation to the speed of the body at the instant it hits the ground.

Interval of time in seconds.

1	$\frac{1}{2}$	$\frac{1}{4}$	$\frac{1}{8}$	$\frac{1}{16}$	$\frac{1}{32}$	$\frac{1}{64}$	$\frac{1}{160}$	$\frac{1}{1600}$	$\frac{1}{16000}$

Distance covered in feet.

144	76	39	$19\frac{3}{4}$	$9\frac{15}{16}$	$4\frac{63}{64}$	$2\frac{127}{256}$	$\frac{1599}{1600}$	$\frac{15999}{160000}$	$\frac{159999}{16000000}$

Average speed in feet per second.

144	152	156	158	159	$159\frac{1}{2}$	$159\frac{3}{4}$	$159\frac{9}{10}$	$159\frac{99}{100}$	$159\frac{999}{1000}$

These approximations approach a limiting value, 160 feet per second, which is defined as the *instantaneous speed* of the bomb upon striking the ground, or what is the same thing, its rate of change of distance with respect to time at that instant.

We may discuss the same example from an algebraic standpoint. The distance covered by a falling body is given by the function $y = 16x^2$, where y is the distance, and x the time elapsed. From this formula, merely by substituting 5 (seconds) for x, we find that y is equal to 400 (feet). How shall we make use of this formula to find the speed at the end of five seconds? Let us fix our attention upon a short interval of time just before the falling object strikes the ground and the correspondingly short interval of distance traversed in that period of time. We shall call this small interval of time Δx*, and the distance traversed in that period Δy. Knowing the value of Δx, having chosen it arbitrarily, the problem is to find the value of Δy. At the beginning of the space interval, Δy, the exact elapsed time since the falling body left the airship was $(5 - \Delta x)$ seconds. The

---

*Read "delta x," not "delta times x"; for Δ is merely a symbol, a direction for performing a certain operation, to wit, taking a small portion of x.

distance covered in the time $(5 - \Delta x)$ seconds is $(400 - \Delta y)$ feet. Our functional relation indicates that

$$\text{Distance} = 16(\text{Elapsed Time})^2.$$

Thus, for the entire fall

$$400 = 16(5)^2,$$

and for the incompleted journey

$$(400 - \Delta y) = 16(5 - \Delta x)^2.$$

This may be simplified to

$$400 - 16(5 - \Delta x)^2 = \Delta y$$
$$400 - 16(25 - 10\Delta x + \Delta x^2) = \Delta y$$
$$400 - 400 + 160\Delta x - 16\Delta x^2 = \Delta y$$
$$160\Delta x - 16\Delta x^2 = \Delta y.$$

The last equation gives the distance Δy in terms of Δx units. To find the *average* speed during the entire time interval Δx, we must form the fraction

$$\text{Average Speed} = \frac{\text{Distance Interval}}{\text{Time Interval}}$$

or

$$\text{Average Speed} = \frac{\Delta y}{\Delta x} = \frac{160\Delta x - 16\Delta x^2}{\Delta x}.$$

Thus,

$$\frac{\Delta y}{\Delta x} = 160 - 16\Delta x.$$

Now as the interval of time Δx is made smaller, that is, as we take closer and closer approximations to the speed at the instant the body strikes the ground (5 seconds having elapsed) the limit of the ratio $\Delta y/\Delta x (= 160 - 16 \Delta x)$ is 160. In other words, as Δx approaches zero in value, the function of Δx (the expression $160 - 16\,\Delta x$) approaches 160. Thus, the *instantaneous speed* at the end of five seconds is 160 feet per second. We indicate that the ratio $\Delta y/\Delta x$ approaches a limit by writing its limiting value as dy/dx. In technical terms

$$\lim_{\Delta x \to 0} \frac{\Delta y}{\Delta x} = \frac{dy}{dx}$$

which may be read, "The limit of $\Delta y/\Delta x$ as Δx approaches zero is dy/dx."

Let us pause for a moment to get our bearings. What have we accomplished? It may seem trivial that with all the elaborate machinery at our disposal we have succeeded only in ascertaining the instantaneous speed of a falling body as it strikes the earth. Yet if our accomplishment is trivial, then motion is trivial as well, for we have, whether we realized it or not, trapped Zeno's arrow in its flight and established the changelessness of our universe. With the aid of the concepts of limit and function, we have made meaningful the notion of change and rate of change. *Change is a functional table.* As an item (independent variable) on one side of the table varies, its corresponding item (dependent variable) on the other side shows a correlative variation. The quotient of change, i.e., the limiting ratio of the two variations, is denoted by rate of change. All the vagaries, the mysteries, and uncertainties indissolubly linked with the idea of motion, are thus swept away or, more appropriately, transformed into a few precise and definable aspects of the idea of function. The limit of a function is exemplified quite simply by the ratio $\Delta y/\Delta x$ as Δx approaches zero. It is easy to see that $\Delta y/\Delta x$ is a function of Δx, in other words, that this ratio is a function of the independent variable Δx. As we assign arbitrary values to Δx, its dependent variable, Δy, assumes a corresponding set of values, and as we have seen, that ratio approaches a limit. It follows that we have not only revealed the meaning of the limit of function but have already made practical use of this concept.

It is now possible to define the fundamental process of the differential calculus, computing the limit of a function, or what is the same thing, determining its derivative. For, in effect, the rate of change of a function is itself a function of that function, and in getting at the limit of the rate of change, the derivative, we are getting at the heart of the machinery of our primitive function.

Assume we wish to determine the rate of change of a function $y = f(x)$ at an arbitrary point x_0. The average change in the function $f(x)$ over an interval extending from x_0 to $x_0 + \Delta x$ is the difference in the value of the function $y = f(x)$ at the two end points, x_0 and $x_0 + \Delta x$, divided by the length between these two end points, $(x_0 + \Delta x) - x_0$. Thus,

$$y_0 = f(x_0)$$

and

$$y_0 + \Delta y = f(x_0 + \Delta x).$$

Whence a change in a function, from the purely algebraic standpoint, is given by

$$\Delta y = f(x_0 + \Delta x) - f(x_0),$$

and the average rate of change of a function, obtained by dividing the change, Δy, by the length of the interval over which that change is taken, Δx, is

$$\frac{\Delta y}{\Delta x} = \frac{f(x_0 + \Delta x) - f(x_0)}{\Delta x}.$$

In order to obtain better approximations to the instantaneous rate of change at the point x_0, it is only necessary to take smaller intervals, that is, to let Δx approach zero.

As Δx approaches zero, the expression $\dfrac{f(x_0 + \Delta x) - f(x_0)}{\Delta x}$

approximates as closely as may be desired to the instantaneous rate of change at x_0. Thus, in the *limit* as Δx approaches zero, the quotient $\dfrac{f(x_0 + \Delta x) - f(x_0)}{\Delta x}$

approaches a limiting value, denoted by dy/dx. It is this which is called the derivative of the function $f(x)$ at the point x_0. But since x_0 is an arbitrary point, the derivative may be said to represent the instantaneous rate of change of a function as the independent variable ranges through an entire set of values.

For the sake of clarity, a geometric interpretation of the derivative may be helpful. Chronologically, the geometric . interpretation preceded the analytic. One of the outstanding problems of the seventeenth century was that of drawing the tangent to a curve at an arbitrary point. It was solved by Newton's predecessor and teacher at Cambridge, Isaac Barrow. On the basis of the geometrical researches of Barrow, Newton developed the concept of the rate of change along analytic lines. The close connection between algebra and geometry, epitomized by the fact that every equation has a graph and every graph an equation, thus bore fruit once more. In the Cartesian plane, let the graph of the function $y = f(x)$ be the curve in Figure III-18.

Consider the points P_1 and P_2 on this curve; their x coordinates are denoted by x_0 and $x_0 + \Delta x$, where Δx is the distance between the projection of the two points on the x axis. The y coordinates of the points P_1 and P_2 are then determined from the equation of the curve and are $f(x_0)$ and $f(x_0 + \Delta x)$ respectively. The slope of the line joining P_1 and P_2 (the tangent of the angle θ) is precisely the quotient

$$\frac{f(x_0 + \Delta x) - f(x_0)}{\Delta x}$$

As we let Δx approach zero, the point P_2 is carried along the curve so that it approaches the point P_1, and the slope of the line (the above quotient) approaches as its limiting

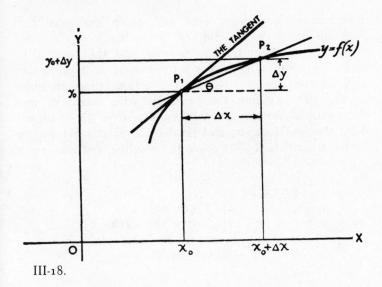

III-18.

value the slope of the tangent to the curve at the point P_1. But the slope of the tangent at that point is numerically

equal to $\dfrac{dy}{dx}$ $\left(\text{since } \underset{\Delta x \to 0}{\text{limit}} \ \dfrac{\Delta y}{\Delta x} = \dfrac{dy}{dx} \right)$. In other words, the

slope of the tangent at every point along a curve is identical with the derivative at that point. Or, to put it differently, the slope of the tangent to a curve gives the direction the curve is taking (i.e., whether it is rising or falling), and thus its rate of change. Thus, the geometric equivalent of the derivative is the slope of the tangent.

We may now recall our statement that the values for which a function attains its maximum or minimum correspond to the points on the curve at which the tangent is horizontal. The slope of a horizontal line is, of course, zero. Since the derivative is identical with the tangent, we may conclude that the maximum and minimum values of a function are those for which the derivative of the function is equal to zero. Many interesting problems can be solved in this way.

The previously discussed problem of determining the

rectangle with greatest area and given perimeter falls into this category. One side of the rectangle was denoted by x, the adjacent side by $2 - x$, and the area, y, by $x(2 - x)$. Since the area is a function of x, its derivative will be equal to zero when the function attains its maximum value. Finding the rectangle with maximum area by means of the calculus entails these steps: (1) Differentiate the function, i.e., find its derivative; (2) Set the derivative equal to 0; (3) Solve the resulting equation for x.

Step I:

$$y = x(2-x)$$
$$y + \Delta y = (x + \Delta x)(2 - x - \Delta x)$$
$$(y + \Delta y) - y = (x + \Delta x)(2 - x - \Delta x) - x(2 - x)$$
$$\Delta y = 2x - x^2 - x\Delta x + 2\Delta x - x\Delta x - \Delta x^2 - 2x + x^2$$
$$\Delta y = 2\Delta x - 2x\Delta x - \Delta x^2$$
$$\frac{\Delta y}{\Delta x} = 2 - 2x - \Delta x$$

$$\underset{\Delta x \to 0}{\text{Limit}} \frac{\Delta y}{\Delta x} = \frac{dy}{dx}$$

$$\text{and } \frac{dy}{dx} = 2 - 2x$$

Step II:

$$\frac{dy}{dx} = 2 - 2x = 0$$

Step III:

$$2 - 2x = 0$$
$$2 = 2x$$
$$1 = x$$

This checks with the result obtained before without the aid of the calculus: the rectangle of maximum area, with a perimeter of 4, is a square each of whose sides equals 1.

More elaborate examples, drawn from the fields of chemistry, economics, physics, etc., require a greater sophistication with respect to mathematical technique, but not with respect to the ideas involved.

By considering the derivative at every point of the interval over which it is defined, we have seen that the derivative is in turn a function of the independent variable. Differentiation need not stop here, for the derived function may also have a derivative, the second derivative of the original function. The notation for the second derivative of $y = f(x)$ is $\dfrac{d^2y}{dx^2}$. The nth derivative of a function is obtained by differentiating the function n times. Its symbol is $\dfrac{d^ny}{dx^n}$. What do these higher derivatives mean?

Usually it is possible to give to the second derivative a physical and geometrical interpretation. If the function $y = f(x)$ represents the distance covered by a falling body in the time x, the first derivative represents the rate of change of distance, with respect to time. The second derivative is the rate of change of the rate of change of distance with respect to time, and is commonly known as the acceleration of the body. For a falling body, the distance $y = 16x^2$ must be differentiated once to obtain the speed and once again to obtain the acceleration. The mathematical details of both differentiations are:

I]
$$y = 16x^2$$
$$y + \Delta y = 16(x + \Delta x)^2$$
$$(y + \Delta y) - y = 16(x + \Delta x)^2 - 16x^2$$
$$= 16(x^2 + 2x\Delta x + \Delta x^2) - 16x^2$$
$$= 16x^2 + 32x\Delta x + 16\Delta x^2 - 16x^2$$
$$\Delta y = 32x\Delta x + 16\Delta x^2$$
$$\frac{\Delta y}{\Delta x} = 32x + 16\Delta x$$
$$\underset{\Delta x \to 0}{\text{Limit}} \frac{\Delta y}{\Delta x} = \frac{dy}{dx}$$
$$\frac{dy}{dx} = 32x.$$

$$\left(\frac{dy}{dx}\right) = 32x$$

$$\left(\frac{dy}{dx}\right) + \Delta\left(\frac{dy}{dx}\right) = 32\,(x+\Delta x)$$

$$\left(\frac{dy}{dx}\right) + \Delta\left(\frac{dy}{dx}\right) - \left(\frac{dy}{dx}\right) = 32\,(x + \Delta x) - 32x$$

$$\Delta\left(\frac{dy}{dx}\right) = 32\Delta x$$

$$\frac{\Delta\left(\dfrac{dy}{dx}\right)}{\Delta x} = 32$$

$$\text{Limit}_{\Delta x \,\to\, 0} \; \frac{\Delta\left(\dfrac{dy}{dx}\right)}{\Delta x} = \frac{d^2y}{dx^2}$$

$$\frac{d^2y}{dx^2} = 32$$

The second derivative is a constant, the number 32. This constant is called the gravitational constant of a falling body due to the earth's gravitational pull. It expresses the remarkable fact that any body, regardless of its mass, dropped from a height 16 feet above the earth (and neglecting air resistance), will strike it in one second, moving at a speed of 32 feet per second at the instant of impact.

So far as the geometric interpretation of the second derivative goes: For curves drawn in the plane, at every point the curvature is directly proportional to the second derivative. To determine the curvature of a given arc, draw the circle which best fits that arc.

The radius of that circle is the radius of curvature, and its reciprocal the curvature.

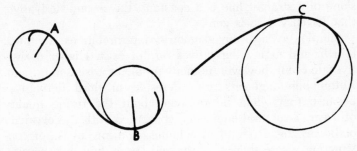

III-19.

Let us see how this is applied, for example, to the straight line. The curvature of a straight line is zero. Any function, the graph of which is a straight line, has an equation of the form $y = mx + b$, where m and b are constants.

Differentiating gives $dy/dx = m$. When m is differentiated, its rate of change or derivative equals zero, since m is a constant. Thus, the first derivative tells us that the

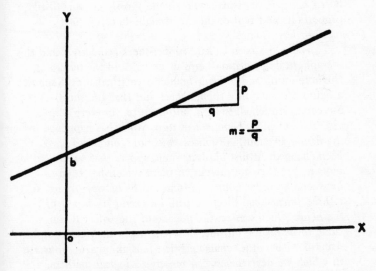

III-20. *The graph of the equation* y = mx + b.

slope of a straight line is a constant; the second derivative, that its curvature is zero.

Simple physical or geometric interpretations of third, fourth, and higher derivatives do not exist. Higher derivatives do occur, however, in many problems arising in physics. Automobile engineers are interested in third derivatives because they yield information about the riding quality of a car. Structural engineers, concerned with the elasticity of beams, the strength of columns, and any phase of construction where there is shear and stress, find first, second, third, and fourth derivatives indispensable; and there exist innumerable other examples in the fields of the physical sciences and statistical applications to the social sciences.

Topology is that branch of geometry which deals with those properties of figures which are unchanged by continual deformations. Henri Poincaré commented about its propositions, "They would remain true if the figures were copied by an inexpert draftsman who should grossly change all the proportions and replace all the straight lines by lines more or less sinuous."

Topology has been called "rubber-sheet geometry," and this metaphorical appellation helps us understand the nature of this interesting and difficult branch of mathematics. Suppose a figure is drawn on a rubber sheet and that the sheet is thereupon stretched, twisted, and distorted in every conceivable way except by being cut or torn. What has happened to the figure? Obviously distances, directions, and sizes have been changed. At first blush it would appear that no similarities remain. Yet they are present. A point which has been between two other points continues to be between them. In a three-dimensional sheet—a ball, for example—a part of the figure which was on the outside of the ball will continue to be on the outside, distort as we will. More complex structures have other characteristics. It is an amazing universe in which we nevertheless can preserve a logical mathematical system. The science, as Gamow points out, examines not

only spheres that have been flattened out, but also such right-
and left-hand objects as gloves and golf clubs, as well as
automobile tires, which are obviously different from football
bladders, pretzels, the various kinds of knots, and even an
astronomical universe which may be closed in on itself. It
deals with a fantastic world, a kind of mathematical puzzle.
Indeed it had its origins in such puzzles as that of the
seven bridges of Königsberg—whether one could travel
over all seven without repeating any part of the journey—
which Euler was able to answer in the negative.

We have stated that modern mathematics is composed
of many different divisions. Despite its rigorousness,
topology is one of the most appealing. Descartes and
Euler made early observations about it. A German mathematician, Listig, examined it systematically in the nineteenth
century. Its study is today one of the largest and most
important of mathematical activities.

Richard Courant was born in Germany and studied at
Breslau, Zurich, and Göttingen. He has done notable work
in the theory of functions and the calculus of variations, and
collaborated with David Hilbert in writing Methoden der
Mathematischen Physik. He taught on the Continent and in
England before joining the staff of New York University,
where he was head of the Department of Mathematics until
his retirement in 1958. His collaborator, Herbert Robbins,
specializes in the theory of probability and mathematical
statistics, and is Chairman of the Department of Mathematical
Statistics at Columbia.

TOPOLOGY

RICHARD COURANT AND HERBERT ROBBINS

Euler's Formula for Polyhedra

ALTHOUGH THE STUDY OF POLYHEDRA held a central place in Greek geometry, it remained for Descartes and Euler to discover the following fact: In a simple polyhedron let V denote the number of vertices, E the number of edges, and F the number of faces; then always

1] $\quad V - E + F = 2$

By a polyhedron is meant a solid whose surface consists of a number of polygonal faces. In the case of the regular solids, all the polygons are congruent and all the angles at vertices are equal. A polyhedron is *simple* if there are no "holes" in it, so that its surface can be deformed continuously into the surface of a sphere. Figure III-22 shows a simple polyhedron which is not regular, while Figure III-23 shows a polyhedron which is not simple.

The reader should check the fact that Euler's formula holds for the simple polyhedra of Figures III-21 and III-22, but does not hold for the polyhedron of Figure III-23.

To prove Euler's formula, let us imagine the given simple polyhedron to be hollow, with a surface made of thin rubber. Then if we cut out one of the faces of the hollow polyhedron, we can deform the remaining surface until it stretches out flat on a plane. Of course, the areas of the faces and the angles between the edges of the polyhedron will have changed in this process. But the network of vertices and edges in the plane will contain the same number of vertices and edges as did the original polyhedron, while the number of polygons will be one less than in the original

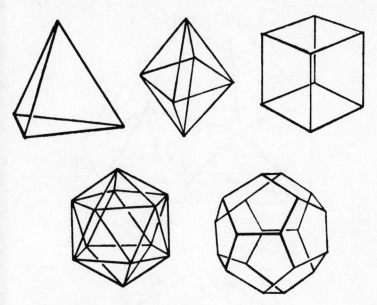

III-21.   *The regular polyhedra.*

polyhedron, since one face was removed. We shall now show that for the plane network, $V - E + F = 1$, so that, if the removed face is counted, the result is $V - E + F = 2$ for the original polyhedron.

First we "triangulate" the plane network in the following way: In some polygon of the network which is not already a triangle we draw a diagonal. The effect of this is to increase both E and F by 1, thus preserving the value of $V - E + F$. We continue drawing diagonals joining pairs of points (Figure III-24) until the figure consists entirely of triangles, as it must eventually. In the triangulated network, $V - E + F$ has the value that it had before the division into triangles, since the drawing of diagonals has not changed it. Some of the triangles have edges on the boundary of the plane network. Of these some, such as *ABC*, have only one edge on the boundary, while other triangles may have two edges on the boundary. We take any boundary triangle and remove that part of it which does not

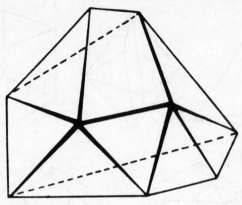

III-22.   A simple polyhedron. $V - E + F = 9 - 18 + 11 = 2$.

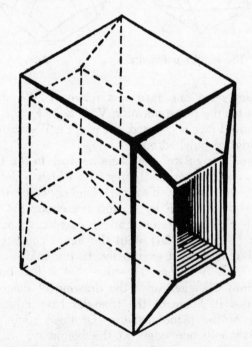

III-23.   A non-simple polyhedron.
$V - E + F = 16 - 32 + 16 = 0$.

also belong to some other triangle. Thus, from ABC we remove the edge AC and the face, leaving the vertices A, B, C and the two edges AB and BC; while from DEF we remove the face, the two edges DF and FE, and the vertex F. The removal of a triangle of type ABC decreases E and F by 1, while V is unaffected, so that $V - E + F$ remains the same. The removal of a triangle of type DEF decreases V by 1, E by 2, and F by 1, so that $V - E + F$ again remains the same. By a properly chosen sequence of these operations we can remove triangles with edges on the boundary (which changes with each removal), until finally only one triangle remains, with its three edges, three vertices, and one face. For this simple network, $V - E + F = 3 - 3 + 1 = 1$. But we have seen that by

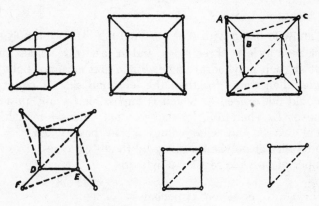

III-24. *Proof of Euler's theorem.*

constantly erasing triangles $V - E + F$ was not altered. Therefore in the original plane network $V - E + F$ must equal 1 also, and thus equals 1 for the polyhedron with one face missing. We conclude that $V - E + F = 2$ for the complete polyhedron. This completes the proof of Euler's formula.

On the basis of Euler's formula it is easy to show that there are no more than five regular polyhedra. For suppose that a regular polyhedron has F faces, each of which is an

n-sided regular polygon, and that r edges meet at each vertex. Counting edges by faces and vertices, we see that

2] $nF = 2E;$

for each edge belongs to two faces, and hence is counted twice in the product nF; moreover,

3] $rV = 2E,$

since each edge has two vertices. Hence from 1] we obtain the equation

$$\frac{2E}{n} + \frac{2E}{r} - E = 2$$

or

4] $$\frac{1}{n} + \frac{1}{r} = \frac{1}{2} + \frac{1}{E}.$$

We know to begin with that $n >$ and $r > 3$, since a polygon must have at least three sides, and at least three sides must meet at each polyhedral angle. But n and r cannot both be *greater* than three, for then the left-hand side of equation 4] could not exceed $\frac{1}{2}$, which is impossible for any positive value of E. Therefore, let us see what values r may have when $n = 3$, and what values n may have when $r = 3$. The totality of polyhedra given by these two cases gives the number of possible regular polyhedra.

For $n = 3$, equation 4] becomes

$$\frac{1}{r} - \frac{1}{6} = \frac{1}{E};$$

r can thus equal 3, 4, or 5. (6, or any greater number, is obviously excluded, since $1/E$ is always positive.) For these values of n and r we get $E = 6$, 12, or 30, corresponding respectively to the tetrahedron, octahedron, and icosahedron. Likewise, for $r = 3$ we obtain the equation

$$\frac{1}{n} - \frac{1}{6} = \frac{1}{E},$$

from which it follows that $n = 3, 4$, or 5, and $E = 6, 12$, or 30, respectively. These values correspond respectively to the tetrahedron, cube, and dodecahedron. Substituting these values for n, r, and E in equations 2] and 3], we obtain the numbers of vertices and faces in the corresponding polyhedra.

Topological Properties of Figures

TOPOLOGICAL PROPERTIES. We have proved that the Euler formula holds for any simple polyhedron. But the range of validity of this formula goes far beyond the polyhedra of elementary geometry, with their flat faces and straight edges; the proof just given would apply equally well to a simple polyhedron with curved faces and edges, or to any subdivision of the surface of a sphere into regions bounded by curved arcs. Moreover, if we imagine the surface of the polyhedron or of the sphere to be made out of a thin sheet of rubber, the Euler formula will still hold if the surface is deformed by bending and stretching the rubber into any other shape, so long as the rubber is not torn in the process. For the formula is concerned only with the *numbers* of the vertices, edges, and faces, and not with lengths, areas, straightness, cross-ratios, or any of the usual concepts of elementary or projective geometry.

We recall that elementary geometry deals with the magnitudes (length, angle, and area) that are unchanged by the rigid motions, while projective geometry deals with the concepts (point, line, incidence, and cross-ratio) which are unchanged by the still larger group of projective transformations. But the rigid motions and the projections are both very special cases of what are called topological transformations: a topological transformation of one geometrical figure A into another figure A' is given by any correspondence

$$p \;\leftrightarrow\; p'$$

between the points p of A and the points p' of A' which has the following two properties:

1] *The correspondence is biunique.* This means that to each point p of A corresponds just one point p' of A', and conversely.

2] *The correspondence is continuous in both directions.* This means that if we take any two points p, q of A and move p so that the distance between it and q approaches zero, then the distance between the corresponding points p', q' of A' will also approach zero, and conversely.

Any property of a geometrical figure A that holds as well for every figure into which A may be transformed by a topological transformation is called a topological property of A, and topology is the branch of geometry which deals only with the topological properties of figures. Imagine a figure to be copied "free-hand" by a conscientious but inexpert draftsman who makes straight lines curved and alters angles, distances and areas; then, although the metric and projective properties of the original figure would be lost, its topological properties would remain the same.

The most intuitive examples of general topological transformations are the deformations. Imagine a figure such as a

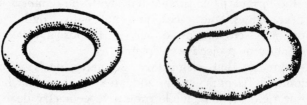

III-25. *Topologically equivalent surfaces.*

III-26. *Topologically non-equivalent surfaces.*

sphere or a triangle to be made from or drawn upon a thin
sheet of rubber, which is then stretched and twisted in any
manner without tearing it and without bringing distinct
points into actual coincidence. (Bringing distinct points into
coincidence would violate condition 1. Tearing the sheet
of rubber would violate condition 2, since two points of the
original figure which tend toward coincidence from opposite
sides of a line along which the sheet is torn would not tend
toward coincidence in the torn figure.) The final position
of the figure will then be a topological image of the original.
A triangle can be deformed into any other triangle or into
a circle or an ellipse, and hence these figures have exactly
the same topological properties. But one cannot deform a
circle into a line segment, nor the surface of a sphere into
the surface of an inner tube.

The general concept of topological transformation is
wider than the concept of deformation. For example, if a
figure is cut during a deformation and the edges of the cut
sewn together after the deformation in exactly the same way
as before, the process still defines a topological transformation of the original figure, although it is not a deformation.
Thus the two curves of Figure III-32 are topologically
equivalent to each other or to a circle, since they may be
cut, untwisted, and the cut sewn up. But it is impossible to
deform one curve into the other or into a circle without first
cutting the curve.

Topological properties of figures (such as are given by
Euler's theorem and others to be discussed in this section)
are of the greatest interest and importance in many mathematical investigations. They are in a sense the deepest and
most fundamental of all geometrical properties, since they
persist under the most drastic changes of shape.

CONNECTIVITY. As another example of two figures that are
not topologically equivalent we may consider the plane
domains of Figure III-27. The first of these consists of all
points interior to a circle, while the second consists of all
points contained between two concentric circles. Any closed
curve lying in the domain *a* can be continuously deformed or

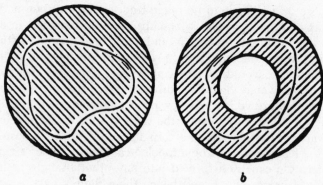

III-27. *Simple and double connectivity.*

"shrunk" down to a single point within the domain. A domain with this property is said to be simply connected. The domain *b* is not simply connected. For example, a circle concentric with the two boundary circles and midway between them cannot be shrunk to a single point within the domain, since during this process the curve would necessarily pass over the center of the circles, which is not a point of the domain. A domain which is not simply connected is said to be multiply connected. If the multiply connected

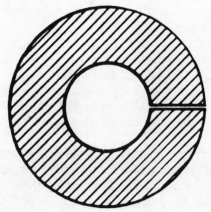

III-28. *Cutting a doubly connected domain to make it simply connected.*

domain *b* is cut along a radius, as in Figure III-28, the resulting domain is simply connected.

More generally, we can construct domains with two, three, or more "holes," such as the domain of Figure III-29. In order to convert this domain into a simply connected domain, two cuts are necessary. If $n - 1$ nonintersecting

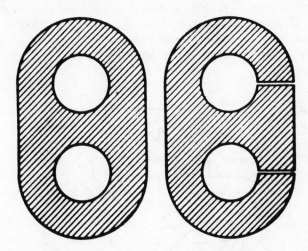

III-29. *Reduction of a triply connected domain.*

cuts from boundary to boundary are needed to convert a given multiply connected domain *D* into a simply connected domain, the domain *D* is said to be *n*-tuply connected. The degree of connectivity of a domain in the plane is an important topological invariant of the domain.

Other Examples of Topological Theorems

THE JORDAN CURVE THEOREM. A simple closed curve (one that does not intersect itself) is drawn in the plane. What property of this figure persists even if the plane is regarded as a sheet of rubber that can be deformed in any way? The length of the curve and the area that it encloses can be changed by a deformation. But there is a topological property of the configuration which is so simple that it may seem trivial: A simple closed curve C in the plane divides the plane into exactly two domains, an inside and an outside. By this is meant that the points of the plane fall into two classes—A, the outside of the curve, and B, the inside —such that any pair of points of the same class can be joined by a curve which does not cross C, while any curve joining a pair of points belonging to different classes must cross C. This statement is obviously true for a circle or an ellipse, but the self-evidence fades a little if one contemplates a complicated curve like the twisted polygon in Figure III-30.

III-30. *Which points of the plane are inside this polygon?*

This theorem was first stated by Camille Jordan (1838– 1922) in his famous *Cours d'Analyse,* from which a whole generation of mathematicians learned the modern concept of rigor in analysis. Strangely enough, the proof given by Jordan was neither short nor simple, and the surprise was even greater when it turned out that Jordan's proof was invalid and that considerable effort was necessary to fill the gaps in his reasoning. The first rigorous proofs of the theorem were quite complicated and hard to understand, even for many well-trained mathematicians. Only recently have comparatively simple proofs been found. One reason for the difficulty lies in the generality of the concept of "simple closed curve," which is not restricted to the class of polygons or "smooth" curves, but includes all curves which are topological images of a circle. On the other hand, many concepts such as "inside," "outside," etc., which are so clear to the intuition, must be made precise before a rigorous proof is possible. It is of the highest theoretical importance to analyze such concepts in their fullest generality, and much of modern topology is devoted to this task. But one should never forget that in the great majority of cases that arise from the study of concrete geometrical phenomena it is quite beside the point to work with concepts whose extreme generality creates unnecessary difficulties. As a matter of fact, the Jordan curve theorem is quite simple to prove for the reasonably well-behaved curves, such as polygons or curves with continuously turning tangents, which occur in most important problems.

THE FOUR-COLOR PROBLEM. From the example of the Jordan curve theorem one might suppose that topology is concerned with providing rigorous proofs for the sort of obvious assertions that no sane person would doubt. On the contrary, there are many topological questions, some of them quite simple in form, to which the intuition gives no satisfactory answer. An example of this kind is the renowned "four-color problem."

In coloring a geographical map it is customary to give different colors to any two countries that have a portion of their boundary in common. It has been found empirically that any map, no matter how many countries it contains nor how they are situated, can be so colored by using only *four* different colors. It is easy to see that no smaller number of colors will suffice for all cases. Figure III-31 shows an island in the sea that certainly cannot be properly colored

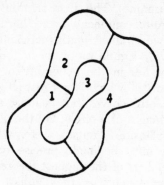

III-31. *Coloring a map.*

with less than four colors, since it contains four countries, each of which touches the other three.

The fact that no map has yet been found whose coloring requires more than four colors suggests the following mathematical theorem: For any subdivision of the plane into non-overlapping regions, it is always possible to mark the regions with one of the numbers 1, 2, 3, 4 in such a way that no two adjacent regions receive the same number. By "adjacent" regions we mean regions with a whole segment of boundary in common; two regions which meet at a single point only or at a finite number of points (such as the states of Colorado and Arizona) will not be called adjacent, since no confusion would arise if they were colored with the same color.

The problem of proving this theorem seems to have been first proposed by Moebius in 1840, later by DeMorgan in 1850, and again by Cayley in 1878. A "proof" was published by Kempe in 1879, but in 1890 Heawood found an error in Kempe's reasoning. By a revision of Kempe's proof, Heawood was able to show that five colors are always sufficient. Despite the efforts of many famous mathematicians, the matter essentially rests with this more modest result: It has been *proved* that five colors suffice for all maps and it is *conjectured* that four will likewise suffice, But, as in the case of the famous Fermat theorem neither a proof of this conjecture nor an example contradicting it has been produced, and it remains one of the great unsolved problems in mathematics. The four-color theorem has indeed been proved for all maps containing less than thirty-eight regions. In view of this fact it appears that even if the general theorem is false it cannot be disproved by any very simple example.

In the four-color problem the maps may be drawn either in the plane or on the surface of a sphere. The two cases are equivalent: any map on the sphere may be represented on the plane by boring a small hole through the interior of one of the regions A and deforming the resulting surface until it is flat, as in the proof of Euler's theorem. The resulting map in the plane will be that of an "island" consisting of the remaining regions, surrounded by a "sea" consisting of the region A. Conversely, by a reversal of this process, any map in the plane may be represented on the sphere. We may therefore confine ourselves to maps on the sphere. Furthermore, since deformations of the regions and their boundary lines do not affect the problem, we may suppose that the boundary of each region is a simple closed polygon composed of circular arcs. Even thus "regularized," the problem remains unsolved; the difficulties here, unlike those involved in the Jordan curve theorem, do not reside in the generality of the concepts of region and curve.

A remarkable fact connected with the four-color problem is that for surfaces more complicated than the plane or the sphere the corresponding theorems have actually been proved, so that, paradoxically enough, the analysis of more complicated geometrical surfaces appears in this respect to be easier than that of the simplest cases. For example, on the surface of a torus (see Figure III-25), whose shape is that of a doughnut or an inflated inner tube, it has been shown that any map may be colored by using seven colors, while maps may be constructed containing seven regions, each of which touches the other six.

KNOTS. As a final example it may be pointed out that the study of knots presents difficult mathematical problems of a topological character. A knot is formed by first looping and interlacing a piece of string and then joining the ends together. The resulting closed curve represents a geometrical figure that remains essentially the same even if it is deformed by pulling or twisting without breaking the string. But how is it possible to give an intrinsic characterization that will distinguish a knotted closed curve in space from an unknotted curve such as the circle? The answer is by no means simple, and still less so is the complete mathematical analysis of the various kinds of knots and the differences between them. Even for the simplest case this has proved to be a sizable task. Consider the two trefoil knots shown in Figure III-32. These two knots are completely symmetrical "mirror images" of one another, and are topologically equivalent, but they are not congruent. The problem arises whether it is possible to deform one of these knots into the other in a continuous way. The answer is in the negative, but the proof of this fact requires considerably more knowledge of the technique of topology and group theory than can be presented here.

III-32. *Topologically equivalent knots that are not deformable into one another.*

The Topological Classification of Surfaces

THE GENUS OF A SURFACE. Many simple but important topological facts arise in the study of two-dimensional surfaces. For example, let us compare the surface of a sphere with that of a torus. It is clear from Figure III-33 that the two surfaces differ in a fundamental way: on the sphere, as in the plane, every simple closed curve such as C separates the surface into two parts. But on the torus there exist closed curves such as C' that do not separate the surface into two parts. To say that C separates the sphere into two parts means that if the sphere is cut along C it will fall into two distinct and unconnected pieces, or, what amounts to the same thing, that we can find two points on the sphere such that any curve on the sphere which joins them must intersect C. On the other hand, if the torus is cut along the closed

 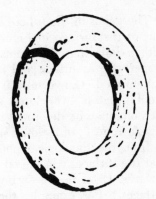

III-33.   *Cuts on sphere and torus.*

curve C', the resulting surface still hangs together: any point of the surface ran be joined to any other point by a curve that does not intersect C'. This difference between the sphere and the torus marks the two types of surfaces as topologically distinct, and shows that it is impossible to deform one into the other in a continuous way.

Next let us consider the surface with two holes shown in Figure III-34. On this surface we can draw two non-intersecting closed curves A and B which do not separate the surface. The torus is always separated into two parts by any two such curves. On the other hand, three closed non-intersecting curves always separate the surface with two holes.

These facts suggest that we define the *genus* of a surface as the largest number of non-intersecting simple closed curves that can be drawn on the surface without separating

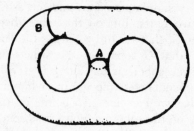

III-34. *A surface of genus 2.*

it. The genus of the sphere is 0, that of the torus is 1, while that of the surface in Figure III-34 is 2. A similar surface with p holes has the genus p. The genus is a topological property of a surface and remains the same if the surface is deformed. Conversely, it may be shown (we omit the proof) that if two closed surfaces have the same genus, then one may be deformed into the other, so that the genus $p = 0, 1, 2 \ldots$ of a closed surface characterizes it completely from the topological point of view. (We are assuming that the surfaces considered are ordinary "two-sided" closed surfaces. Later in this section we shall consider "one-sided" surfaces.) For example, the two-holed doughnut and the sphere with two "handles" of Figure III-35 are both closed surfaces of genus 2, and it is clear that either of these sur-

III-35. *Surfaces of genus 2.*

faces may be continuously deformed into the other. Since the doughnut with p holes, or its equivalent, the sphere with p handles, is of genus p, we may take either of these surfaces as the topological representative of all closed surfaces of genus p.

THE EULER CHARACTERISTIC OF A SURFACE. Suppose that a

closed surface S of genus p is divided into a number of regions by marking a number of vertices on S and joining them by curved arcs. We shall show that

1] $$V - E + F = 2 - 2p,$$

where V = number of vertices, E = number of arcs, and F = number of regions. The number $2 - 2p$ is called the *Euler characteristic* of the surface. We have already seen that for the sphere, $V - E + F = 2$, which agrees with 1], since $p = 0$ for the sphere.

To prove the general formula 1], we may assume that S is a sphere with p handles. For, as we have stated, any surface of genus p may be continuously deformed into such a surface, and during this deformation the numbers $V - E + F$ and $2 - 2p$ will not change. We shall choose the deformation so as to ensure that the closed curves A_1, A_2, B_1, B_2 . . . where the handles join the sphere consist of arcs of the given subdivision. (We refer to Figure III-36, which illustrates the proof for the case $p = 2$.)

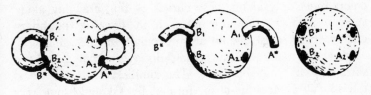

III-36.

Now let us cut the surface S along the curves A_2, B_2 . . . and straighten the handles out. Each handle will have a free edge bounded by a new curve A^*, B^* . . . with the same number of vertices and arcs as A_2, B_2 . . . respectively. Hence $V - E + F$ will not change, since the additional vertices exactly counterbalance the additional arcs, while no new regions are created. Next, we deform the surface by flattening out the projecting handles, until the resulting surface is simply a sphere from which $2p$ regions have been removed. Since $V - E + F$ is known to equal 2 for any subdivision of the whole sphere, we have

$$V - E + F = 2 - 2p$$

for the sphere with $2p$ regions removed, and hence for the original sphere with p handles, as was to be proved.

Figure III-23 illustrates the application of formula 1] to a surface S consisting of flat polygons. This surface may be continuously deformed into a torus, so that the genus p is 1 and $2 - 2p = 2 - 2 = 0$. As predicted by formula 1],

$$V - E + F = 16 - 32 + 16 = 0.$$

ONE-SIDED SURFACES. An ordinary surface has two sides. This applies both to closed surfaces like the sphere or the torus and to surfaces with boundary curves, such as the disk or a torus from which a piece has been removed. The two sides of such a surface could be painted with different colors to distinguish them. If the surface is closed, the two colors never meet. If the surface has boundary curves, the two colors meet only along these curves. A bug crawling along such a surface and prevented from crossing boundary curves, if any exist, would always remain on the same side.

Moebius made the surprising discovery that there are surfaces with only *one* side. The simplest such surface is the so-called Moebius strip, formed by taking a long rectangular strip of paper and pasting its two ends together after giving one a half-twist, as in Figure III-37. A bug crawling along this surface, keeping always to the middle of the strip, will return to its original position upside down (Figure III-38). Anyone who contracts to paint one side of a Moebius strip could do it just as well by dipping the whole strip into a bucket of paint.

Another curious property of the Moebius strip is that it has only one edge, for its boundary consists of a single closed curve. The ordinary two-sided surface formed by pasting together the two ends of a rectangle without twisting has two distinct boundary curves. If the latter strip is cut along the center line it falls apart into two different strips of the same kind. But if the Moebius strip is cut along this line (shown in Figure III-37) we find that it remains in one piece. It is rare for anyone not familiar with the Moebius strip to predict this behavior, so contrary to one's intuition of what "should" occur. If the surface that results from cutting the Moebius strip along the middle is again cut along its middle, two separate but intertwined strips are formed.

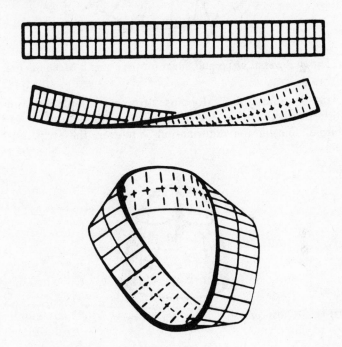

III-37. *Forming a Moebius strip.*

It is fascinating to play with such strips by cutting them along lines parallel to a boundary curve and $\frac{1}{2}$, $\frac{1}{3}$, etc., of the distance across. The Moebius strip certainly deserves a place in elementary geometrical instruction.

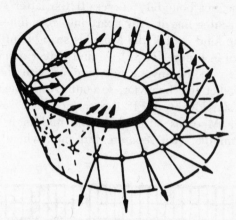

III-38.   *Reversal of up and down on traversing a Moebius strip.*

The boundary of a Moebius strip is a simple and un-knotted closed curve, and it is possible to deform it into a circle. During the deformation, however, the strip must

III-39.   *Cross-cap.*

be allowed to intersect itself. (Hence, such a deformation of a "real" paper Moebius strip is only possible in the imagination.) The resulting self-intersecting and one-sided surface is known as a cross-cap (Figure III-39). The line of intersection *RS* is regarded as two different lines, each belonging to one of the two portions of the surface which intersect there. The one-sidedness of the Moebius strip is preserved because this property is topological; a one-sided surface cannot be continuously deformed into a two-sided surface.

Another interesting one-sided surface is the "Klein bottle." This surface is closed, but it has no inside or outside. It is topologically equivalent to a pair of cross-caps with their boundaries coinciding.

III-40.   *Klein bottle.*

It may be shown that any closed, *one-sided* surface of genus $p = 1, 2 \ldots$ is topologically equivalent to a sphere from which p disks have been removed and replaced by cross-caps. From this it easily follows that the Euler characteristic $V - E + F$ of such a surface is related to p by the equation

$$V - E + F = 2 - p.$$

The proof is analogous to that for two-sided surfaces. First we show that the Euler characteristic of a cross-cap or Moebius strip is 0. To do this we observe that, by cutting across a Moebius strip which has been subdivided into a number of regions, we obtain a rectangle that contains two more vertices, one more edge, and the same number of regions as the Moebius strip. For the rectangle, $V - E + F = 1$, as we proved on pages 209–211. Hence for the Moebius strip $V - E + F = 0$. As an exercise, the reader may complete the proof.

It is considerably simpler to study the topological nature of surfaces such as these by means of plane polygons with certain pairs of edges conceptually identified. In the diagrams of Figure III-41, parallel arrows are to be brought in-

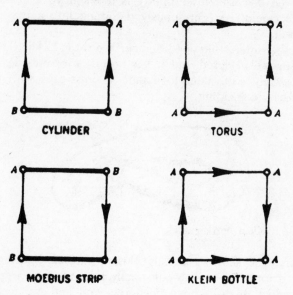

III-41. Closed surfaces defined by coordination of edges in plane figure.

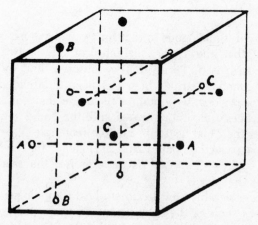

III-42. Three-dimensional torus defined by boundary identification.

to coincidence—actual or conceptual—in position and direction.

This method of identification may also be used to define three-dimensional closed manifolds, analogous to the two-dimensional closed surfaces. For example, if we identify corresponding points of opposite faces of a cube (Figure III-42), we obtain a closed, three-dimensional manifold called the three-dimensional torus. This manifold is topologically equivalent to the space between two concentric torus surfaces, one inside the other, in which corresponding points of the two torus surfaces are identified (Figure III-43). For the latter manifold is obtained from the cube if two pairs of conceptually identified faces are brought together.

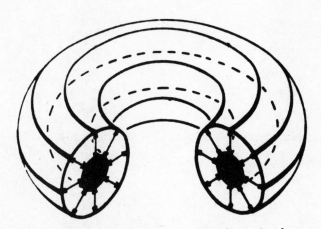

III-43. *Another representation of three-dimensional torus. (Figure cut to show identification.)*

IV. MATHEMATICS and the World Around Us

In 1937, a group of mathematicians on the staff of Washington Square College, New York University, collaborated in writing an Introduction to Mathematics which, in revised form, continues to have marked success as "a survey emphasizing mathematical ideas and their relations to other fields of knowledge." Unlike many similar texts, it lays strong emphasis on the cultural aspects of mathematics as it applies to "the sciences, the arts, philosophy, and to knowledge in general." The techniques of mathematics are to be found in the volume, but it is the conviction of the authors that "mathematics has much more to offer . . . than mere training in memorizing formulas and manipulating symbols." This view coincides with that which has been taken in the preparation of the present volume, and makes the following selection particularly pertinent. One of the authors, Hollis R. Cooley, Professor of Mathematics at Washington Square College, is a member of the Board of Academic Editorial Advisers of the Library of Science, of which the present volume is a part.

MATHEMATICS AND MODERN CIVILIZATION

HOLLIS R. COOLEY,
DAVID GANS, MORRIS KLINE, AND
HOWARD E. WAHLERT

THERE ARE TWO MAJOR WAYS in which mathematics has become so effective in our age. The first is through its relationships with science. Because the distinguishing characteristic of modern civilization is the extent to which the physical sciences have molded it, this relationship of mathematics to science bears directly on the importance of mathematics in modern civilization. The second way in which mathematics has materially influenced our age has been through its connection with human reasoning. Mathematical method is reasoning on the highest level known to man, and every field of investigation, be it law, politics, psychology, medicine, or anthropology, has felt its influence and modeled itself on mathematics to some extent. This chapter will elaborate on these two major contributions of mathematics.

The Relation of Mathematics to the Sciences

In order to gain a more comprehensive view of the relation of mathematics to the sciences, let us analyze the various ways in which mathematics serves scientific investigation. A] *Mathematics supplies a language for the treatment of the quantitative problems of the physical and social sciences.* Much of this language takes the form of mathematical symbols. Workable rules for carrying out operations are made possible by symbols; without them the simple operations of arithmetic would be extremely clumsy and the solution of even simple equations would be very difficult.

Symbols also permit concise, unambiguous representation of ideas which are sometimes quite complex. Consider, for example, how much is involved in the calculus symbol, Dy. Once the meaning and use of a symbol has been grasped, there is no need to think through the origin and development of the idea symbolized, each time it is used. One of the chief reasons that mathematics has been so effective in problems that have been insoluble by other methods is that it has powerful techniques based upon the use of symbols. Scientists have learned to use mathematical symbols whenever possible.

B] *Mathematics supplies science with numerous methods and conclusions.* Among the important conclusions which mathematics furnishes are its formulas, which scientists accept and use in solving problems. The use of such formulas is so common that the contribution of mathematics in this direction is not always appreciated. A brief survey will show that this contribution is an important one. Consider, for example, the many geometric formulas, such as those for areas and volumes, the formulas for finding the distance and velocity of moving objects, and formulas for compound interest and annuities. The realization of the importance of such formulas would be further strengthened if we could live for a time in the world as it was before many of the problems of science were solved by mathematical means.

Over and above the content of mathematics are methods which mathematics has developed and which are advantageously employed by scientists. Indirect measurement by means of the trigonometric ratios is an example of such a method, as is the representation of a curve by means of an equation for the purpose of studying the curve. The calculus likewise presents the sciences with a method for finding rates of change of varying quantities, as well as other methods. Of course, in a larger sense all of mathematics is a method, but this aspect of mathematics will be discussed in the latter part of the chapter.

Any discussion of the methodological contribution of mathematics to science cannot overlook the fact that mathematics is often an essential part of scientific method. The scientist observes or experiments so as to discover new facts and, more important, so as to secure a basis for a theory or hypothesis which should explain a large class of phenomena. Often this hypothesis either takes the form of a mathematical law, or includes mathematical laws. Mathematical methods are very frequently used to deduce consequences of the laws. Further observation or experimentation checks these deductions and thereby tests the suitability of the hypothesis.

C] *Mathematics enables the sciences to make predictions.* This is perhaps the most valuable contribution of mathematics to the sciences. If a structural engineer wishes to use a beam of given size at the base of a skyscraper, he wants to know in advance whether or not it is able to carry the load to which it will be subjected. It would not do for him to use the beam only to find out that it is too weak when the building collapses. In order to determine whether the beam is strong enough he uses mathematical analysis. If machines are to be built, an industrialist wants to know beforehand what the machines will cost and what savings in the cost of production they will effect. The answers to these questions are obtained largely by the use of mathematical methods. The accurate determination of the standard of living of a country, comprising such factors as wages and living costs, is made by means of statistical analyses. The occurrence of an eclipse is far less astonishing than man's ability to predict the time of its occurrence. The same can be said for predictions in many sciences.

The ability to make predictions by mathematical means was exemplified in a most remarkable way in 1846 by two astronomers, Leverrier and Adams. Each of these men, working independently, decided, as a result of his calculations, that there must exist another planet beyond those known at the time. A search for it in the sky at the predicted place and time revealed the planet which we now call Neptune. When one considers that Neptune is not visible to the naked eye and can be found with a telescope only with exact knowledge of its position, the extraordinary precision of this prediction is apparent. It should be realized that prediction plays a part in every mathematical solution of a quantitative problem arising in the physical and social sciences.

D] *Mathematics supplies science with ideas with which to describe phenomena.* Among such ideas which mathematics has furnished for science may be mentioned the idea of a functional relation; the graphical representation of functional relations by means of coordinate geometry; the notion of a limit, which provides methods of determining instantaneous velocity and acceleration and of calculating areas and volumes; and the notion of infinite classes, which helps us, among other things, to understand motion. Of special importance are the statistical methods and theories which mathematics furnishes to science and which have led to the idea of a statistical law. It may be worthwhile to call attention to the important fact that by the use of the notion of a statistical law, scientists are enabled to reason about situations which appear to be utterly chaotic. The seemingly random motions of the molecules of a liquid, like the seemingly random motions of people on a crowded thoroughfare, may be found in some cases to obey a law as closely as does the earth in its motion around the sun.

A description of the extent to which mathematical concepts are used by scientists is not complete without mention of the fact that for many physical phenomena no exact concepts exist other than mathematical ones. The physicist uses extensively the concept of voltage or electromotive force which causes electric current to flow in wires. The layman makes use of this voltage when he connects his lamp or radio to a socket in a room of his home. Yet there is no precise physical concept of voltage nor a good physical explanation of what causes current to flow in wires. There are, however, numerous exact mathematical relationships which involve the concept of voltage, as well as others equally vague physically. For example, Ohm's law, which states that the voltage between any two points on a wire is the product of the current flowing and the resistance in that wire, in such an exact relationship involving voltage. Another physical concept which can be represented and discussed only in mathematical terms is the notion of an ether wave which carries light, radio broadcasts, X-rays, and other electromagnetic phenomena. It is very significant to recall in this connection that Clerk Maxwell predicted the existence of a wide class of these waves (that light was a wave motion in ether was accepted long before Maxwell's time) only because he had a mathematical term in his equations which should have some physical significance. No physical understanding of ether waves exists even today. Nevertheless the marvels of radio broadcasting and television are realized because we can discuss and work with this concept mathematically.

The impress of mathematics on our civilization through the sciences is nowhere better exemplified than by the work of Maxwell. One has only to contemplate the number of people who are entertained, educated, and propagandized by radio broadcasts to realize the magnitude of the effect.

The effect of television broadcasting staggers the imagination.

E] *Mathematics has been of use to science in preparing men's minds for new ways of thinking.* The concepts of importance in science today, elementary though some of them may seem, came to men with great difficulty. The concepts of a force of gravity, of energy, and of limitless space, took years to develop, and genius was required to express them precisely. Many times in the history of science, progress was possible only because mathematical thinking led the way. For example, by A.D. 1600 algebra had developed to such an extent that an algebraic expression, like $x^2 + 2x + 5$, suggested the idea of a functional relation between the variable x and a variable y, formed by setting $y = x^2 + 2x + 5$. Moreover, the great number of algebraic expressions called attention to the variety of possible relations between variables. Soon after 1600 scientists began to attack their problems by seeking the mathematical relationships between variables, and they used algebraic expressions to represent these relationships. Descartes, Galileo, Huygens, Leibnitz, and Newton were prominent among the discoverers of such relationships. Of course there were other factors in seventeenth-century civilization which stimulated scientific activity, but much of this activity would have been impossible without the aid of the idea of a function. Among the functions perfected by 1600 were the trigonometric functions, which play an important part in the study of problems involving periodic motion. In the seventeenth century, Galileo studied the periodic motion of the pendulum, Newton discovered that sound is produced by a periodic motion of air molecules, and Huygens advanced the theory that light was a periodic wave motion.

A more recent example of the way in which mathematical thought has prepared men's minds for new developments in science is furnished by the change in the conception of space. This took place to a great extent during the middle of the nineteenth century, and was produced by discoveries of systems of non-Euclidean geometry. Early in the present century Minkowski emphasized the necessity of linking time and space in order to have a proper understanding of the way in which physical events take place. Still later, Einstein recast some of the most fundamental physical notions by utilizing the mathematical ideas of non-Euclidean geometry and of space-time as a single concept. Great as is the genius of Einstein, it is almost certain that he was able to achieve some of his results only because mathematicians of preceding decades had suggested new ways of thinking about space and time.

The purpose of this article has been to indicate broadly several ways in which mathematics is useful to science. To summarize: mathematics supplies a language, methods, and conclusions for science, enables scientists to predict results, furnishes science with ideas for describing phenomena, and prepares the minds of scientists for new ways of thinking.

It would be quite wrong to think that mathematics gives so much to the sciences and receives nothing in return. The physical objects and quantities with which the sciences work and the observed facts concerning those objects and quantities often serve as a source of the elements and postulates of mathematics. It is true that the elements and postulates thus suggested need not be used, and that a mathematical system can be based on elements and postulates which have no apparent application in the physical world. But this does not alter the fact that actually the fundamental concepts of many branches of mathematics are the ones suggested by physical experiences. This statement is especially true of Euclidean geometry. On a more advanced level the result of the Michelson-Morley experiment in physics was adopted as an important basic assumption in the theory of relativity.

A further service rendered by science to mathematics is found in the fact that scientific theories frequently suggest directions for pursuing mathematical investigation and thus furnish a starting point for mathematical discoveries. For example, Copernican astronomy suggested many new problems involving the effects of gravitational attraction between heavenly bodies in motion. These problems stimulated activity in the field of differential equations.

The Mathematization of Science

The point of the preceding section is that mathematics is useful to science in many ways. But that fact in itself fails to describe completely the relation of mathematics to science. A further fact of importance in this connection is that science is becoming more and more mathematical in its concepts and in its methods. Many fundamental physical concepts are being replaced by mathematical ones. Since the time of Newton physicists have believed confidently in the existence of a force called gravity, which causes objects to attract each other. As a consequence of the theory of relativity, the concept of a force of gravity is replaced by an essentially mathematical concept, namely, that the nature of space is such as to cause objects to move as they do, just as the form of a railroad track determines the path of a train running on it.

Science is tending to become more mathematical not only in its concepts, but also in its methods. For example, science now uses abstract concepts far removed from observed physical substances. These concepts are fictions introduced to form a theory. One such fiction is the substance called ether. No physical evidence showed the existence of this substance, but some medium was considered necessary for the transmission of light, much as air serves as a medium in the transmission of sound; hence, the existence of ether with definite properties was assumed. Surely no concept of mathematics is less "real" than ether. Its position in physics is analogous to the position of an undefined term in mathematics, such as a line. We know nothing about a mathematical line except those properties which the postulates imply, and similarly we know nothing about ether except those properties which it is assumed to have.

The assumption of the existence of abstract elements in science is not limited to ether. Gravity, we found, is in the same class. In addition, in the study of atomic structure today, science assumes the existence of entities such as electrons, protons, positrons, neutrons, and others. The only justification for the assumption of their existence is that they help to simplify the explanation of observed facts. They permit the development of a logical theory of atomic structure. But they are not entities which we can observe directly. That physics and chemistry, which avowedly attempt to explain the world of our sense experience, should resort to these fictions or ideal elements is far more surprising than that mathematics, which acknowledges its abstract nature, should do so. Yet, such is the state of affairs in modern science.

Finally, science has become more mathematical in its greater and greater reliance on deductive reasoning as a means of arriving at truth. This tendency is understandable. The certainty of conclusions obtained by deductive reasoning from accepted facts of experience is preferable to the uncertainty of conclusions gained by experiment and generalization therefrom. Moreover, the use of abstract concepts in science requires the use of deductive reasoning, because one cannot experiment with abstractions as with tangible elements.

One may ask what postulates are at the basis of deductive reasoning in physics, for example. Experience suggests to physicists statements which seem to apply to the world about us. These statements are taken as postulates and are used as a basis for reasoning. The so-called law of the conservation of energy is an example of a physical postulate. It is a common observation that when energy is used to do work, other energy appears. If muscular energy is used in sawing wood, energy in the form of heat raises the temperature of the wood and the saw. The energy latent in coal is used to give energy in the form of electricity. From these, and numerous other examples, many physicists have been willing to accept as axiomatic the statement that in a physical or chemical process energy is never lost, but may be converted into a new form. Newton's laws of motion are further postulates for physics. Of course, in so far as the properties of space are involved in their work, physicists use the postulates about space and the consequent deductions from the postulates made by mathematicians.

To sum up the ways in which science has become mathematical, we may state that science has tended to replace physical concepts by concepts which are mathematical in nature, that it has adopted the use of abstract concepts and postulates to explain phenomena, and, finally, that it has become more mathematical by making greater use of the deductive method of reasoning.

The philosopher Kant once remarked that the degree of development of science depends upon the extent to which it has become mathematized. By this criterion the physical sciences have reached a high degree of development today. But even physics is not entirely mathematical, and there are branches of chemistry and biology in which the use of mathematics is of minor importance, as it is in most social sciences. Even in some of those fields in which mathematics is little employed, however, there are many who believe that if mathematical ideas and methods were used more extensively, progress would be more rapid.

The philosopher Descartes devoted himself to the problem of finding a method of obtaining truths which should be applicable to all fields. He finally decided that mathematics supplied the answer to his problem. Applied to any one field of investigation, the method consists essentially of selecting certain basic concepts, securing facts or relations involving these concepts about which one could be certain, and then deducing conclusions from these fundamental facts. To Descartes the existence of mathematics was proof that the method worked, and he proceeded to apply it to philosophy.

Descartes was both right and wrong. He was wrong in supposing that all mathematical conclusions are truths. The creation of non-Euclidean geometry has taught us to be wary of any statements proclaimed as unquestionable truths. But Descartes was right in his choice of mathematical method as an outstanding method of obtaining useful conclusions.

The method Descartes urged should, in the light of our present knowledge of mathematics, be stated thus: Choose concepts that appear to be basic and accept, as postulates, statements about these concepts that appear to be supported by experience. Deduce conclusions by strict reasoning. The conclusions will then be as certain as the postulates. In addition to the above one might well add: Employ symbols and quantitative and geometrical relationships where possible. Thereby numerous specialized mathematical methods and conclusions will become applicable.

Mathematical method is something over and above the mere use of mathematical formulas and conclusions. It is an approach to problems which may be employed in almost all fields. The physical sciences have consciously employed it for centuries, and the biological and social sciences are doing so more and more.

Let us see how the social sciences use mathematical method. We shall consider the school of economic thought known as the single tax system and proposed by Henry George. From several principles which George believed to be basic in our economic system he sought to deduce further economic laws. The desirability of the society to which these conclusions pointed caused George to urge strongly the acceptance of his principles.

Land, says Henry George, is the basis of all wealth. All taxation should be on land itself but not on improvements on land such as buildings. Moreover, the tax on land should be large enough to discourage the ownership of unimproved land. On the other hand, taxes on labor or the products of industry discourage industry.

From these assumptions Henry George and his followers deduced many interesting conclusions. For example, the high tax rate on unimproved land would make it unprofitable to own land without utilizing it. An owner would therefore build or farm on his land. This step would result in considerable industry and in farming, thereby creating a market for labor and preventing unemployment, and producing goods for consumption. Moreover, since labor would be essential to the utilization of land either as a farm or as an industrial site, labor would be valued and properly rewarded. Involuntary poverty would be abolished.

Whether or not this economic system, which still has many supporters, is attractive to the reader, it does illustrate how one may approach economic problems by means of mathematical method. While this school of thought was chosen because its postulates and reasoning are readily stated and understood, it is not unique in its use of mathematical method.

Philosophers, especially, have been aware of mathematical method and of its value in attempting to arrive at truths. The student who would care to sample the entertaining arguments contained in Plato's dialogues will see example after example of deductive reasoning carried out on the basis of statements and definitions initially accepted by the disputants. He would also enjoy the surprising and sometimes startling conclusions to which the reasoning leads.

Some philosophers have gone further in their use of mathematical method. The philosophers Thomas Aquinas and Baruch Spinoza were not satisfied merely to use the essentials of this method. To insure the accuracy of their reasoning they stated explicity the postulates and theorems of their respective systems of philosophy, and proved each theorem carefully by reference to previous deductions or to the postulates. While the use of this mathematical style of stating theorems and their proofs breeds a stilted literary style, these philosophers were seeking truth and preferred to sacrifice style in order to guarantee the accuracy of their conclusions.

It may be said of almost all philosophers that they seek to employ deductive reasoning from postulates acceptable to them. Sometimes, it must be admitted, assumptions creep into their reasoning which were not intended to be part of their systems.

Mathematical method has impressed the logicians themselves. They, too, have analyzed human reasoning to find the basic principles of reasoning acceptable to all men. From these basic principles or postulates, "laws" (theorems) of reasoning are deduced, to which all acceptable reasoning must conform. The logicians have gone so far as to develop an extensive symbolism which aids them just as symbols aid mathematicians. Every modern textbook on logic now teaches this symbolic logic.

Knowledge of the nature of mathematical method pays dividends to the average man who seeks to understand and cope with political, religious, and economic problems. When carefully analyzed, different political doctrines, as well as religious and economic doctrines, differ essentially in the postulates on which they are based. Statements acceptable to some people as fundamental truths are regarded by others as unacceptable and sometimes unreasonable assumptions. It follows, therefore, that the conclusions correctly deduced from such postulates will not be equally acceptable to all people.

A glance at current economic theories illustrates the remarks of the preceding paragraph. The differences between social and economic systems, such as socialism and capitalism, might well be reduced to differences in fundamental assumptions concerning the acquisition and ownership of wealth. Shall natural resources such as coal and waterpower be the property of a few people or of the whole population? Shall profits be unlimited or should the tax rate be larger for corporations with higher profits so as to prevent excessive profits? Is the contribution of men's labor to a business an investment as is money, or is labor to be paid for as a commodity, on the basis of the supply and demand? Does the government have the obligation to employ people who are not employed by industry and, if so, can it tax to secure money to pay those people? Such fundamental issues are at the heart of economic systems. Once a person commits himself to one or another side of issues like these, the whole body of his economic beliefs follows as a consequence. Much dispute would be avoided if people would recognize the importance of unearthing the fundamental assumptions on which differing economic beliefs are based and concentrate their discussion on these assumptions.

A person's decision to adopt one or another set of basic economic assumptions is entirely analogous to the scientist's decision to adopt one or another system of geometry. This analogy goes further. When scientists found that a non-Euclidean geometry fitted observations and experience better than Euclidean geometry, the latter was rejected and the former installed in its place. A revolution took place in scientific thought. The same happens in economic thought. Individuals and sometimes nations find that an economic system does not meet the test of experience, that is, does not meet the needs of the people. Individuals react by changing their economic beliefs. Nations sometimes react by revolution, for often the economic system is tied to the political system or imposed by a ruling group.

In political systems, too, basic assumptions determine entire theories. Before each presidential-election campaign in this country the politicians dare to be logical. The leading members of each party gather to draw up the party's platform. This platform contains the basic principles to which party members supposedly adhere. These principles are, in a real sense, postulates. From them the party's position on public issues should be deducible. Needless to say the usual party platform uses numerous undefined terms. Needless to say, also, a party's actual position on public issues does not always follow logically from the postulates contained in the platform. And frequently the postulates are surreptitiously changed after the campaign gets under way.

Mathematics and the Culture of a Civilization

The influence of mathematics on our civilization through the medium of the sciences, and the direct influence of mathematics on all fields of thought as a major method of attacking problems, are the larger values of the subject. These values together with the uncountable applications and relationships of mathematics to engineering, art, philosophy, music, logic, religion, and the social sciences, establish mathematics as having unchallengeable importance for our civilization and for the student of our civilization. It will be noted that the importance of mathematics extends beyond the ways in which man earns his daily bread. It includes those higher forms of human activity such as art, philosophy, and music, which are commonly referred to as the cultural fields.

Some students of the history of culture, in particular the late Oswald Spengler, go further in their judgment of the relation of mathematics to the culture of a civilization. For example, Spengler maintains that a study of the mathematics of any civilization will reveal characteristics which are common to other forms of expression of the culture of that civilization. By other forms of expression is meant literature, painting, music, architecture, science, philosophy, and the like.* It is not maintained that this correspondence extends to every detail of a culture but that it is typical of the culture. We shall illustrate this thesis, but must warn the reader that the point of view is not necessarily held by all competent writers on the history of civilizations.

Historians such as Spengler emphasize the following characteristics of Greek mathematics. First, it was mainly static; it dealt with the properties of figures at rest in space. This is illustrated by the geometry of Euclid. Second, their mathematics was for the most part confined to bounded figures, lying in small regions of space, for example, the circle and the triangle. Although the Greeks did regard a straight line as being infinite in extent, and defined parallel lines to be lines which do not meet however far extended, they did not carry far the idea of a geometrical infinity, nor did they study other concepts of infinity. Third, Greek mathematics was confined to objects which could be visualized, that is, to geometry. Although Pythagoras, in his study of length, discovered irrational numbers in the sixth century B.C., the Greeks never developed the abstract idea of irrational numbers (irrational numbers were studied as line segments), nor, for that matter, did they use or develop algebra to any extent. Negative numbers, zero, and imaginary numbers were unknown to them. Fourth, form in mathematics was valued, as the emphasis on deductive reasoning in geometry shows.

*A discussion of these interrelationships, which indicates agreement with Spengler's views, will be found in *Science and the Human Temperament* by Erwin Schrödinger, the famous atomic physicist.

Compare these characteristics of Greek mathematics with the following facts about Greek life and thought. Greek architecture, as revealed in their temples, was, to a large extent, static, that is, it did not suggest the idea of motion. To many people their temples present an appearance of repose, of being well balanced and firmly set on the ground. The type of physics studied by the Greeks was the branch of mechanics now known as static, a study of the forces acting on bodies at rest. Just as their geometry was confined to bounded figures, so in their lives the Greeks were relatively unexplorative. They lived in city-states and, as compared with other great nations, stayed near home. They are accused of a lack of perspective, or the representation of depth, in painting. Their music is called two-dimensional by some writers because it consisted only of rhythm and melody. Harmony was to come later. Their preference for mathematical concepts that can be visualized is reflected in their high development of simple forms in architecture and sculpture.

Let us now make a similar, though partial, comparison of the mathematics and culture of the civilization commencing with the Renaissance, roughly the fourteenth century. The development of algebra by the Hindus and others increased the ability of the mathematician to solve quantitative problems. The notion of a variable, implicit even in the elementary formulas of the Egyptians and the Greeks, became significant. With that, came the idea of a function or relation among variables. Next, the question of rate of change of a function was raised. The attempts to answer this question produced the idea of a limit and its application to functions. Mathematics had become dynamic and was concerned with change. Another characteristic of modern mathematics is the freedom it enjoys to develop concepts having no obvious counterparts in the physical world. This tendency became significant with the rise of non-Euclidean geometry. The realization by mathematicians that they were free to develop any system they chose, regardless of whether or not they started with postulates taken from experience, encouraged them to explore new fields. They constructed a

logical theory to answer the bothersome questions about
infinity, and they began to study spaces which had, at the
time of their investigation, no correspondence to the physical
world.

Let us compare these characteristics of mathematics with
the following developments in other fields. Science began
in the sixteenth century to make a quantitative study of the
world. Physical laws appeared in functional form. Velocity
and acceleration, which are nothing more than rates of
change of functions, became basic objects of study. A science
of dynamics, that is, the study of motion, arose and has
become very important. The earth was explored and the
heavens were studied, thus enlarging our knowledge of
space. Even Gothic architecture, with its tall buildings and
spires, is regarded by some authorities as a sign of ex-
plorative tendency. Science has recently adopted logically
constructed descriptions of phenomena which appeal to the
mind rather than the senses. Philosophy has become more
concerned with, and influenced by, the results of science.
Modern art likewise reflects appeal to the mind rather than
the eye alone.

This topic could be carried much further, but to do so it
would be necessary to analyze other forms of our culture.
The foregoing paragraphs have for their purpose only to
explain the meaning of the statement that the characteristics
of mathematics are related to the characteristics of the
other forms in which the culture of a civilization expresses
itself. It could be shown, further, by means of a detailed
analysis, that the history of mathematical thought is inter-
related with the history of civilization. This statement does
not imply that mathematics caused the changes which
produced one civilization from another, but merely that it
changed with the civilizations and to a large extent reflected
each civilization.

The greatest distinction between our present western
civilization and all others of which we have any knowledge
is the growth of mathematics and the natural sciences and

the application of these fields to industry, engineering, and commerce.

Without belittling the merits of our historians, economists, philosophers, writers, poets, painters, and statesmen, it is possible to say that other civilizations have produced their equals, not merely in ability but in accomplishments. On the other hand, though Euclid and Archimedes were undoubtedly great and though our mathematicians and scientists were able to reach higher, only because, as Newton put it, they were able to stand on the shoulders of such giants, nevertheless it is in our age that mathematics and science have attained their maturity and extraordinary applicability.

Because mathematics has left its imprint upon so many aspects of present-day civilization, its position in the modern world is a fundamental one, and a knowledge of mathematics is essential for a comprehensive understanding of current life and thought.

Eric Temple Bell is the author not only of Men of Mathematics *in which the biography of Gauss reprinted on page 29 appeared, but also of two ingratiating companion volumes entitled* The Queen of the Sciences *and* The Handmaiden of the Sciences. *The former discusses pure mathematics, the latter mathematics in its applications. The oldest of these applications, and in many ways the most striking, was to astronomy. With their relatively undeveloped methods—both astronomical and mathematical—the ancients were nevertheless able to make astonishingly good guesses about the size of the earth, the distance of the sun, and the dates of occurrence of eclipses. The greatest glory which these sciences share is the Newtonian concept of the universe, which Copernicus, Kepler, and other predecessors of Newton helped originate. How astonishing it was that even the stars in their courses obeyed laws which were rigidly mathematical! How almost unbelievable it must have seemed*

*that the planets as they moved around the sun followed
paths described by one of the conic sections—the ellipse!
Educated moderns are so accustomed to these ideas that they
find it almost impossible to share in the wonder which the
first observers must have felt when they learned that our universe is mathematical. And with each new discovery, with
the contemporary theories of a four-dimensional universe
and of an expanding universe, it becomes ever more apparent
that "God is a mathematician."*

FROM CYZICUS TO NEPTUNE

ERIC TEMPLE BELL

A Royal Road

TO ONE OF Alexander the Great's tutors, the Greek
mathematician Menaechmus, is attributed the discouraging
remark "There is no royal road to geometry." Alexander had
impatiently ordered Menaechmus to abridge his proofs.
Unable to oblige his impetuous pupil, Menaechmus nevertheless, perhaps in spite of himself, did succeed in leveling
another royal road. This was the straight highway to the
true beginning of mathematical astronomy and therefore
also of analytical mechanics and mathematical physics. Without the purely mathematical inventions of this somewhat
obscure Greek geometer it is inconceivable that the course
of the physical sciences, in particular mathematical physics
and theoretical astronomy, could have followed even re-
motely any such direction as they actually have.

Of the life of Menaechmus little is known beyond his
problematical dates, 375–325 B.C., and the uncertified
tradition that he succeeded the incomparable Eudoxus
(408–355 B.C.), precursor of the integral calculus, as
director of the mathematical seminar at Cyzicus. Far more

important for science than all the trivialities of Menaechmus'
forgotten life is the memorable fact that he invented the
conic sections. It was the simple geometry of these curves
that led to the beginning of modern astronomy. The conics
are easily visualized.

Imagine a cone standing on a circular base B. The surface of the cone is to be extended (as in the figure)
indefinitely in both directions through the vertex V. The
two parts of this extended cone issuing from V (one up, the
other down), are called nappes; the straight line AVA'
through V perpendicular to B is the axis, and any straight
line, such as G, which passes through V and lies on the
surface is a generator (see figure). The curve of inter-

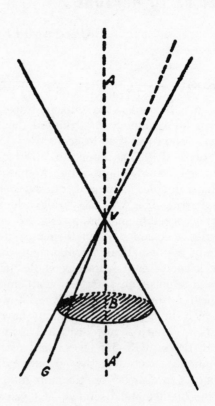

IV-1.

section of a plane with either or both nappes is a conic section, or briefly, a conic. According to this definition it is easily seen that there are precisely seven species of conics.

In 1, 2, 3 the plane passes through V.

1] If the plane passes through V and does not cut the surface elsewhere, the conic is a point.

2] If the plane passes through V and touches the surface, the conic is a straight line (or pair of coincident straight lines).

3] If the plane passes through V and also intersects B in distinct points, the conic is a pair of intersecting straight lines:

In 4, 5, 6, 7 the plane does not pass through V.

4] If the plane cuts the axis at right angles, the conic is a circle.

5] If the plane is not parallel to the axis or to a generator, the conic is an ellipse.

6] If the plane is parallel to a generator, the conic is a parabola.

7] If the plane is parallel to the axis, the conic is a hyperbola, consisting of two branches.

The first two of these are of no interest. Note however that the point conic can be considered as a circle with radius zero, and that the straight line conic is a degenerate case of the pair of intersecting straight lines—when the lines coincide. For what is to follow, the ellipse is the most interesting of the conics, although the parabola also has useful properties, two of which may be noted in passing.

A parabola is approximately the path of a ball, a bullet, or a shell in the air. If the air offered no resistance the path would be exactly a parabola. Thus if warfare were conducted in a vacuum, as it should be, the calculations of ballistics would be much simpler than they actually are, and it would cost considerably less than the $25,000 or so of taxes which it is now necessary to shoot away in order to slaughter one patriot.

Parabolic mirrors offer a somewhat less bloody application of the conics, such mirrors being used in some automobile headlights. Suppose we were required to construct a mirror which would reflect the light from a point-source in a beam of parallel rays. Trial and error might grope for centuries to discover what mathematics reveals with a turn of the hand: there is exactly one type of mirror which will do what is wanted, namely the parabolic. Moreover the calculation prescribes the unique point at which the light must be placed in front of the mirror to produce the parallel rays. This point is called the focus of the parabola.

Passing to the ellipse, which will shortly assume the role of guide in mathematical astronomy, we must define its foci. First, to draw an ellipse, we tie a thread to two pins stuck in the drawing board, say at F and F', and keep the thread taut with the point P of the pencil; P then traces out the curve—an ellipse—permitted by this restraint (see figure).

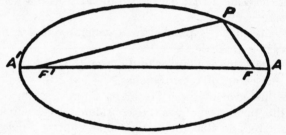

IV-2.

Let the line joining F, F' cut the ellipse in the points A, A'. Then the segment AA' is called the major axis of the ellipse, and each of F, F' is a focus of the ellipse. From the manner in which the curve was drawn it is clear that the sum of the distances, PF, PF' is constant—the same—for all positions of the point P on the ellipse; this sum is equal to AA'.

If the foci F, F' move into coincidence the ellipse degenerates to a circle. Thus, in a sense, an ellipse is a generalization of a circle. This is of some importance as the earlier Greek astronomers chose the too obvious circle as the key to the geometry of the heavens, whereas the ellipse would have fitted more refined observations better. . . .

All the conics are unified from the algebraic point of view of analytic geometry. The several species of conics correspond to the essentially distinct types of equations of the second degree in two variables.

Kepler's Faith

Through the labors of a host of Greek mathematicians, of whom Apollonius (c. 260–200 B.C.) towers up as one of the greatest geometers of all time, the geometry of the conic sections was minutely elaborated long before the decline of Greek learning. There is no evidence that any of the Greek mathematicians suspected that conics would some day prove of paramount importance in the dynamics of the solar system. The contrary appears to be the case. Otherwise Ptolemy (2nd century A.D.) might have been tempted to try ellipses instead of circles as a clue to the geometry of planetary orbits—the Copernican picture of the solar system with all the planets revolving round the Sun is sometimes said to have been imagined (and forgotten) by the Greeks. The Greeks, by the way, seem to have done everything in modern science but never to have done much with it. Possibly they were too vague, too philosophical, to state clearly what, if anything, they really meant.

In devoting the hard labor they did to conic sections the Greek geometers set a fashion followed in modern times by the majority of professional mathematicians. Disregarding any practical or scientific use their creations might have then or later, the pioneers developed mathematics for its own interest. Even from the standpoint of the crassest practicality this austere fashion seems to have been profitable enough. It can be argued, of course, that strict attention to the immediately useful might have proved more profitable, even for mathematics. That, however, is not how science has actually developed, and speculations on how mathematics might have evolved certainly cannot affect the past, although they may inspire educational reformers to new and more frightful excesses.

After the decline of Greece as a leader in geometrical

research, knowledge of the conics was fostered and transmitted to Christian Europeans by sagacious infidels, notably the Arabs. All through this protracted nightmare the Ptolemaic description of the solar system, with the intolerable complexity of its cycles upon cycles, epicycles upon epicycles, brooded like a god. We need not describe Ptolemy's masterpiece. Except for its historical interest this massive work is now happily ignored. The Ptolemaic theory was one of humanity's major blunders in its gropings to touch what it cannot see.

Unfortunately for the progress of knowledge the Ptolemaic theory, with the Earth at the center of everything, was fully competent to account for the observed motions of the planets. This was its damnatory excellence. And more, with sufficient ingenuity the theory could be modified to accommodate certain new observations. Had not the simpler heliocentric picture of the solar system disclosed itself to Copernicus we might still be admiring the Ptolemaic description of the Solar System as one of the sublimest achievements of the human mind. Possibly it was; the Copernican revolution swept it into limbo.

Familiar as we are now with the rapid obsolescence of physical theories, it is difficult for us to imagine the terrific uproar occasioned by Copernicus' fundamental remark that all the planets revolve round the Sun. Nor perhaps did Copernicus himself foresee the full fury of the row his work was to precipitate when, in 1543, he touched on his deathbed the first printed copy of his treatise on the motions of the heavenly bodies. More prudent than his successor Galileo (1564–1642), Copernicus saved himself considerable embarrassment by resigning from life in the nick of time.

The bigoted intelligentsia of all colors, from humanists and theologians to astronomers and mathematicians, rallied to the flag of authority. Unlike the sophisticates of our own generation the sages of the sixteenth century seem to have preferred complexity to simplicity. They would have none of the beautifully direct heliocentric theory. The Chinese involution of the Ptolemaic system, which should have been buried the day Copernicus died, was embalmed and fitted with new gears to preserve it in a ghastly semblance of life for decades after it was dead. Not so many years ago we witnessed a similar brainless rush to the banner of tradition when Einstein amended (and abolished) the Newtonian law of gravitation. Einstein escaped the concentration camps, it is true; yet—we are assured by his former associates and compatriots—his theory is as trivial as was that of Copernicus.

After Copernicus the next long stride toward a rational celestial mechanics was taken by the astronomer Tycho Brahe (1546–1601). Either the laborious Tycho was too absorbed in his observations to have time for mathematics, or he lacked the right type of mind to synthesize his masses of data into a simple geometrical picture. What he had recorded of the motions of the planets, with an accuracy seldom approached before his time, ached for a mathematical interpreter. Did the planets revolve around the Sun in circles, or was some less banal curve demanded to portray their orbits? Johann Kepler (1571–1630), at one time Tycho's assistant, was to find the solution.

A highly gifted mathematician, a terrific worker, a first rate astronomer, and a man of undeviating intellectual honesty, Kepler was the ideal candidate to sift the accumulated data, to calculate endlessly, and to be satisfied with nothing short of the completest accuracy attainable at the time. Undeterred by poverty, failure, domestic tragedy, and persecution, but sustained by his mystical belief in an attainable mathematical harmony and perfection of nature, Kepler persisted for fifteen years before finding the simple regularity he sought.

What Kepler accomplished is one of the most astounding feats of arithmetical divination in the history of science. His labors were rewarded by three of the grandest empirical discoveries ever made, Kepler's laws: all the planets describe ellipses round the Sun, which is at one focus of these ellipses; the line joining the Sun to any planet sweeps over equal areas in equal times; the squares of the periodic times of the planets are proportional to the cubes of the major axes of their orbits. All this he inferred by what was

little better than common arithmetic abbreviated by logarithms.

The story of Kepler's epochal discoveries is so familiar
that we need not dwell on it here, except to note the
curious inconsequence of it all. This applies with equal force
to every mathematical formulation of natural "laws" which
we shall notice later.

What stimulated Kepler to keep slaving all those fifteen
years? An utter absurdity. In addition to his faith in the
mathematical perfectibility of astronomy, Kepler also believed wholeheartedly in astrology. This was nothing against
him. For a scientist of Kepler's generation astrology was as
respectable scientifically and mathematically as the quantum
theory or relativity is to theoretical physicists today. Nonsense now, astrology was not nonsense in the sixteenth
century. . . .

Calculation Plus Insight

The full power of the mathematical method is not displayed in any such laborious cut-and-try as Kepler sweated
over to find his three laws. The hidden strength of mathematics partly reveals itself at the next and far more difficult encounter with the unknown.

Before Kepler's laws could be deduced from something
simpler a mathematics of continuous change had to be invented. This was the differential calculus (described in an
earlier chapter), brought to a usable state of perfection in
the seventeenth century by Newton (1642–1727) and
Leibnitz (1646–1716). In addition the empirical laws of
motion had to be discovered and stated in a form adapted
to mathematical reasoning. Galileo (1564–1642) and Newton between them accomplished this. Finally, although it
may not have been logically required (as will be seen
when we consider Einstein's work), the concept of *force*
had to be clarified, in particular the notion of a force of
attraction between material bodies such as the Sun and the
planets. This was accomplished in Newton's law of universal
gravitation. With these partly empirical preliminaries dis-

posed of the solar system was ready for idealization into a map of reality of sufficient abstractness to be explored mathematically.

The laws mentioned above may be recalled. First, the three laws of motion:

1] Every body will continue in its state of rest or of uniform motion in a straight line unless it is compelled to change that state by impressed force.

2] Rate of change of motion is proportional to the impressed force, and takes place in the direction in which the force acts.

3] Action and reaction are equal and opposite.

The first of these defines inertia; the second (in which "motion" means momentum, or "mass times velocity," both mass and velocity being measured in the appropriate units), introduces the intuitive notion of a *rate*. The last is probably the most important contribution of mathematics to science. . . .

Newton's law of universal gravitation has a more audible mathematical ring:

4] Any two particles of matter in the universe attract one another with a force which is proportional to the product of the mass of the particles, and inversely proportional to the square of the distance between them.

Thus, if m, M are numbers measuring the masses of the particles, and d measures the distance between the particles, the force of attraction is measured by

$$k \times \frac{m \times M}{d^2},$$

in which k is some constant number depending only on the units in terms of which mass, distance, and force are measured. If the distance is doubled, the attractive force is only one-fourth of what it was before; if the distance is trebled, the force is diminished to one-ninth, and so on.

It should appear reasonable even to one who remembers no mathematics beyond arithmetic that the path of one particle being attracted by another will depend upon the law of attraction between the two particles. If the attraction is according to the Newtonian law of the inverse square (square=second power) of the distance the path will bear but little resemblance to that compelled by a law of the inverse third, or fourth, or fifth . . . power of the distance. It should also be fairly obvious that the following question admits a definite answer: If a body, say a comet, is observed to trace a certain definitely known path, characterized geometrically, in its approach to the Sun, what law of attraction between the Sun and the comet will account for the observed path?

A similar problem was proposed in August, 1684 to Newton. For some months the astronomer Halley and other friends of Newton had been discussing the problem in the following precise form: What is the path of a body attracted by a force directed toward a fixed point, the force varying in intensity as the inverse square of the distance? Newton answered instantly, "An ellipse." "How do you know?" he was asked. "Why, I have calculated it." Thus originated the imperishable *Principia*, which Newton later wrote out for Halley. It contained a complete treatise on motion.

When solved fully by Newton this problem answered both the direct and inverse forms of the question proposed, and accounted at one stroke for Kepler's three laws. The laws were deduced from the simple law of universal gravitation. The "inverse" answer showed that if the path is an ellipse the law of attraction is the Newtonian.

It is definitely known that the almost Greek methods of the *Principia*, with their rigid geometrical deductions, are not those by which Newton reached his results, but that he used the calculus as an immeasurably more penetrating instrument of exploration. The calculus being a strange and prickly novelty at the time, Newton wisely recast his findings in the classical geometry familiar to his contemporaries. Today the deduction of Kepler's laws from the Newtonian law of gravitation is accomplished by means of the calculus in a page or two in the textbooks on dynamics. In Newton's day it was a task for a titan. We do seem to progress in some things.

Newton's law contained vastly more than Kepler's too smooth laws. Combined with the calculus it made possible an attack on the observed irregularities of the supposedly perfect elliptical orbits. It is clear that if Newton's law is indeed universal, then every planet in the solar system must perturb every other and cause it to depart from the true ellipse which it would follow if it were the only planet in the system and if both it and the Sun were perfect homogeneous spheres. As the mass of the Sun greatly exceeds that of the planets, the observed irregularities will be slight. But they will be none the less important. We shall see next a spectacular example of the importance of attending to slight though awkward discrepancies between oversimplified mathematical perfection and obstinate facts of observation. Newton himself initiated the study of perturbations which was to enlarge our knowledge of the solar system and prove useful in our own day in the yet more difficult field of atomic structure.

Mathematical Prophecy

The next episode in this brief story of the royal road opens with Newton in his great prime and closes with a young man in his early twenties. As only one main road is being followed from the Cyzicus of Menaechmus to the Cambridge of John Couch Adams, we must pass by unexplored the many alluring highways branching off to other great empires of mathematical reasoning, and condense the boundless work of a century to one fleeting glimpse. At the end of the road we shall see a typical example—also one of the most famous—of the prophetic power of mathematics when applied to a masterly abstraction of nature. Newton's law of universal gravitation was such an abstraction; the methods of mathematical analysis developed from the calculus elicited from Newton's law what it implicitly concealed.

Even today we look in vain for any generalization of physical science which has unified any such vast mass of diverse phenomena as was reduced to a coherent unity by

Newton's law. At a first glance we may think that some of the recent generalizations have a scope as wide as Newton's had in its heyday, but a little consideration shows that this is an illusion. Not only did universal gravitation as mathematicized in Newton's law sweep together, sift, and simplify the scattered astronomical knowledge of twenty centuries or more; it subjected the mysterious tides to mathematical rule, and for over two centuries served as a suggestive guide in all fields of physical science where there was any glimmer of hope that a mechanical philosophy could simplify and unify the data of the senses. The extreme simplicity of the law itself is equalled only by one other powerful generalization of modern science, the law of the conservation of energy, which asserts that the total amount of energy in the universe remains constant. Electric energy, for instance, may be transformed into heat and light, and seem to be dissipated, but actually nothing has been lost. This great generalization (now radically modified since the advent of relativity) may justly claim to have been a descendant of Newton's.

In the century following Newton a host of powerful mathematicians, including Euler (1702–1783), Lagrange (1736–1813), and Laplace (1749–1827), explored the heavens with Newton's universal law as their sole guide, finding almost everywhere measurably complete accord between theory and observation. The mathematical methods used by Newton's successors were not those of the *Principia,* but the more flexible analysis which evolved from the first rather crudely stated forms of the differential and integral calculus.

Not all of these great mathematicians believed in the Newtonian law as a truly universal principle. Euler, for one, doubted whether anything so elementary could possibly account for even so simple a situation as that posed by the motion of the Moon. By "simple" here we mean simple only to uninstructed intuition. The motion of the Moon offers one of the most complicated problems in the whole range of dynamical astronomy. So great are the mathematical difficulties that any analyst of Euler's time might well have believed the Newtonian hypothesis of the inverse square law to be inadequate. The Newtonian law solves the problem of two attracting bodies completely and with a Keplerian simplicity. When more than two bodies attract one another according to the Newtonian law of universal gravitation, no exact solution of the problem of completely describing their motions exists even today. The method of successive approximations is applied to yield progressively more accurate descriptions of the motion sufficient for all practical purposes, such as those demanded by the computation of the nautical almanac. The motion of the Moon is a "three-body problem," the bodies being the Earth, the Sun, and the Moon.

Among those who doubted the universality and adequacy of the Newtonian law was G. B. Airy (1801–1892), for long Astronomer Royal of Newton's own England. Airy's doubt was perfectly legitimate, and indeed highly creditable to a man of science, even if based on a total misconception of the nature of the real difficulties involved. But no skeptic has a right to impose the inertia of his disbelief on young men ardently pushing forward to what they believe are attainable discoveries. To Airy more than to any other relic of academic conservatism is due the official indifference with which the calculations of young Adams— recounted in a moment—were received, when he ventured to sustain Newton by one of the most brilliant mathematical predictions in all the brilliant history of mathematical science. We have not space to go into the record of the official lethargy which robbed Adams of a unique "first"; so with this brief (but on the whole adequate) obituary of his chief obstructor we shall pass on to what Adams did.

On March 13, 1781 William Herschel discovered with his telescope a new member of the Sun's family, the planet subsequently known as Uranus (after a happily abortive attempt to name it for King George of England). The newcomer offered a superb opportunity for testing the Newtonian law.

Before long discrepancies between calculation and observation appeared in the orbit of Uranus that could not be explained away by postulating faulty arithmetic or defective telescopes. Newton's hypothesis simply did not fit the observed facts.

But man is a theorizing animal, and the erratic irregularities of Uranus' motion were attributed to the perturbations of some more distant planet, as yet undiscovered. This hypothetical planet, attracting Uranus according to the Newtonian law of the inverse square, would account for everything provided only it existed, and its mass and orbit were of the right mathematical specifications to produce exactly those irregularities in the orbit of Uranus which had actually been observed.

Mathematically the problem was an inverse one of extreme difficulty. It would be laborious but comparatively easy to calculate the effect of a known planet on the motion of Uranus. But given the erratic motion it was much more difficult to reach out with mathematical analysis into the vague of ultra-planetary space and discover how massive, and where, the unobserved perturber was at any particular date.

By the early 1840's many astronomers believed that such an ultra-Uranian planet existed. In a historic prophecy Herschel stated on September 10, 1846, "We see it [the hypothetical planet] as Columbus saw America from the coast of Spain. Its movements have been felt, trembling along the far-reaching line of our analysis with a certainty hardly inferior to that of ocular demonstration."

The analysis to which Herschel referred was modern mathematics, as it then existed, applied to the Newtonian law. Mathematically the problem of Uranus was this: given the perturbations, to discover the mass and the orbit of the unknown planet producing them—in short, to discover the unobserved member of the Sun's family responsible for the indisputable disharmony in the otherwise harmonious Newtonian symphony of the solar system.

No simile can convey the difficulty of such a problem to anyone who has not seen something similar attempted. "A needle in a haystack" might be suggested, but here we do not know whether the haystack exists or, if it did exist, in what country it might be, or whether, after all, there is a needle to be found.

About eleven months before Herschel prophesied, young Adams (1819–1892) had sent to the Astronomer Royal numerical estimates of the mass and orbit of the undiscovered planet. Moreover Adams had calculated where and when the hypothetical planet perturbing Uranus could be observed in telescopes. Subsequent events proved his calculations correct within a reasonable margin of accuracy. Had a patch of sky no larger than three and a half Moons been searched when Adams told the Astronomer Royal to look, the planet would have been found. But Adams at the time was only an unknown quantity of twenty-four. As an undergraduate at Trinity College, Cambridge, he had resolved to attack the Uranian problem the moment he was quit of his examinations. On taking his degree in January 1843 with the highest honors, he immediately set about his self-imposed task.

In the meantime a seasoned mathematical astronomer in France, U. J. J. Leverrier (1811–1877), had also attacked the perturbations of Uranus in an attempt to locate and estimate the unknown perturber. He also succeeded. Both Adams and Leverrier worked in complete ignorance of what the other was doing.

Leverrier was the luckiest in his friends. The policy of "wait and don't see" delayed search for the planet by the English astronomers until Leverrier's livelier continental friends had already located the suspect in the heavens— very approximately where both Adams and Leverrier had instructed practical astronomers to direct their telescopes.

Thus was Neptune discovered by pure mathematical analysis applied to a great physical hypothesis, and thus ended one glorious canto of the epic begun by Menaechmus, carried on by Kepler, and sped well on to its climax —as yet unpredictable—by Newton.

Not only do the stars and the nebulae obey mathematical law; so also do the simplest as well as the most complex members of the biological world. All unconsciously, the Epeira spins its web according to "a whole armoury of scientific formulae." The engineer constructing a bridge or the physicist planning an electronic experiment seem to work no more meticulously. The "beautiful rose window" which it constructs, as Fabre shows, is geometrically accurate in all its details.

For the past generation and for many perceptive readers of today, one of the most delightful introductions to natural science has been the works of Jean Henri Fabre. Born at St. Léons, France in 1823, the son of illiterate peasants, he received a scanty education and embarked on the career of a typical impoverished schoolmaster of his day. But there were in young Fabre a gift for observation and talent for expression which distinguished him from his fellows. Patiently and laboriously, he educated himself in biology, astronomy, and mathematics. Much like Gilbert White, author of the classic The Natural History and Antiquities of Selborne, he used his own environment as a field of observation and study, and in the process became the author of a series of charming works, many of which, like The Life and Love of the Insect, Social Life in the Insect World, and The Life of the Fly (from which the following selection is taken), were translated into English. His Souvenirs Entomologiques was crowned by the Institute of France.

Despite the popularity of his books, he never achieved riches, and in his seventies, tragedy struck. A secondary school for girls had been established in Avignon, and Fabre, always interested in the young, volunteered his services free. He lectured on the physical and natural sciences, but the bigoted townspeople had no use for such instruction—particularly of young girls. The resulting scandal rocked the town, and Fabre, always unpopular because of his abnormal shyness, was forced to resign his professorship and retire to a garden in Provence. Life seemed over; but for fourteen years he continued his studies and produced some of his finest work.

THE GEOMETRY OF THE EPEIRA'S WEB

J. HENRI FABRE

I FIND MYSELF confronted with a subject which is not only highly interesting, but somewhat difficult: not that the subject is obscure; but it presupposes in the reader a certain knowledge of geometry: a strong meat too often neglected. I am not addressing geometricians, who are generally indifferent to questions of instinct, nor entomological collectors, who, as such, take no interest in mathematical theorems; I write for any one with sufficient intelligence to enjoy the lessons which the insect teaches.

What am I to do? To suppress this chapter were to leave out the most remarkable instance of Spider industry; to treat it as it should be treated, that is to say, with the whole armoury of scientific formulæ, would be out of place in these modest pages. Let us take a middle course, avoiding both abstruse truths and complete ignorance.

Let us direct our attention to the nets of the Epeiræ, preferably to those of the Silky Epeira and the Banded Epeira, so plentiful in the autumn, in my part of the country, and so remarkable for their bulk. We shall first observe that the radii are equally spaced; the angles formed by each consecutive pair are of perceptibly equal value; and this in spite of their number, which in the case of the Silky Epeira exceeds two score. We know by what strange means the Spider attains her ends and divides the area wherein the web is to be warped into a large number of equal sectors, a number which is almost invariable in the work of each species. An operation without method, governed, one might imagine, by an irresponsible whim, results in a beautiful rose window worthy of our compasses.

We shall also notice that, in each sector, the various chords, the elements of the spiral windings, are parallel to one another and gradually draw closer together as they

269

near the center. With the two radiating lines that frame them they form obtuse angles on one side and acute angles on the other; and these angles remain constant in the same sector because the chords are parallel.

There is more than this: these same angles, the obtuse as well as the acute, do not alter in value, from one sector to another, at any rate so far as the conscientious eye can judge. Taken as a whole, therefore, the rope-latticed edifice consists of a series of cross-bars intersecting the several radiating lines obliquely at angles of equal value.

By this characteristic we recognize the "logarithmic spiral." Geometricians give this name to the curve which intersects obliquely, at angles of unvarying value, all the straight lines or "radii vectores" radiating from a center called the "pole." The Epeira's construction, therefore, is a series of chords joining the intersections of a logarithmic spiral with a series of radii. It would become merged in this spiral if the number of radii were infinite, for this would reduce the length of the rectilinear elements indefinitely and change this polygonal line into a curve.

To suggest an explanation why this spiral has so greatly exercised the meditations of science, let us confine ourselves for the present to a few statements of which the reader will find the proof in any treatise on higher geometry.

The logarithmic spiral describes an endless number of circuits around its pole, to which it constantly draws nearer without ever being able to reach it. This central point is indefinitely inaccessible at each approaching turn. It is obvious that this property is beyond our sensory scope. Even with the help of the best philosophical instruments, our sight could not follow its interminable windings and would soon abandon the attempt to divide the invisible. It is a volute to which the brain conceives no limits. The trained mind, alone more discerning than our retina, sees clearly that which defies the perceptive faculties of the eye. The Epeira complies to the best of her ability with this law of the endless volute. The spiral revolutions come closer together as they approach the pole. At a given distance they stop abruptly; but, at this point, the auxiliary spiral, which is not destroyed in the central region, takes up the thread; and we see it, not without some surprise, draw nearer to the pole in ever-narrowing and scarcely perceptible circles. There is not, of course, absolute mathematical accuracy, but a very close approximation to that accuracy. The Epeira winds nearer and nearer round her pole so far as her equipment, which like our own, is defective, will allow her. One would believe her to be thoroughly versed in the laws of the spiral.

I will continue to set forth, without explanations, some of the properties of this curious curve. Picture a flexible thread wound round a logarithmic spiral. If we then unwind it, keeping it taut the while, its free extremity will describe a spiral similar at all points to the original. The curve will merely have changed places.

Jacques Bernouilli, to whom geometry owes this magnificent theorem, had engraved on his tomb, as one of his proudest titles to fame, the generating spiral and its double, begotten of the unwinding of the thread. An inscription proclaimed, *"Eadem mutata resurgo:* I rise again like unto myself."* Geometry would find it difficult to better this splendid flight of fancy towards the great problem of the hereafter.

There is another geometrical epitaph no less famous. Cicero, when quæstor in Sicily, searching for the tomb of Archimedes amid the thorns and brambles that cover us with oblivion, recognized it, among the ruins, by the geometrical figure engraved upon the stone: the cylinder circumscribing the sphere. Archimedes, in fact, was the first to know the approximate relation of circumference to diameter; from it he deduced the perimeter and surface of the circle, as well as the surface and volume of the sphere. He showed that the surface and volume of the last-named equal two-thirds of the surface and volume of the circumscribing cylinder. Disdaining all pompous inscription, the learned Syracusan honored himself with his theorem as his sole epitaph. The geometrical figure proclaimed the individual's name as plainly as would any alphabetical characters.

To have done with this part of our subject here is another property of the logarithmic spiral. Roll the curve along an indefinite straight line. Its pole will become displaced while still keeping on one straight line. The endless scroll leads to rectilinear progression, the perpetually varied begets uniformity.

Now is this logarithmic spiral, with its curious properties, merely a conception of the geometers, combining number and extent, at will, so as to imagine a tenebrous abyss wherein to practise their analytical methods afterwards? Is it a mere dream in the night of the intricate, an abstract riddle flung out for our understanding to browse upon?

No, it is a reality in the service of life, a method of construction frequently employed in animal architecture. The Mollusc, in particular, never rolls the winding ramp of the shell without reference to the scientific curve. The firstborn of the species knew it and put it into practice; it was as perfect in the dawn of creation as it can be to-day.

Let us study, in this connection, the Ammonites, those venerable relics of what was once the highest expression of living things, at the time when the solid land was taking shape from the oceanic ooze. Cut and polished lengthwise, the fossil shows a magnificent logarithmic spiral, the general pattern of the dwelling which was a pearl palace, with numerous chambers traversed by a siphuncular corridor.

To this day, the last representative of the Cephalopoda with partitioned shells, the Nautilus of the Southern Seas, remains faithful to the ancient design; it has not improved upon its distant predecessors. It has altered the position of the siphuncle, has placed it in the center instead of leaving it on the back but it still whirls its spiral logarithmically as did the Ammonites in the earliest ages of the world's existence.

And let us not run away with the idea that these princes of the Mollusc tribe have a monopoly of the scientific curve. In the stagnant waters of our grassy ditches, the flat shells, the humble Planorbes, sometimes no bigger than a duckweed, vie with the Ammonite and the Nautilus in matters of higher geometry. At least one of them, *Planorbis vortex*, for example, is a marvel of logarithmic whorls.

In the long-shaped shells, the structure becomes more complex, though remaining subject to the same fundamental laws. I have before my eyes some species of the genus Terebra, from New Caledonia. They are extremely tapering cones, attaining almost nine inches in length. Their surface is smooth and quite plain, without any of the usual ornaments, such as furrows, knots or strings of pearls. The spiral edifice is superb, graced with its own simplicity alone. I count a score of whorls which gradually decrease until they vanish in the delicate point. They are edged with a fine groove.

I take a pencil and draw a rough generating line to this cone; and, relying merely on the evidence of my eyes, which are more or less practised in geometric measurements, I find that the spiral groove intersects this generating line at an angle of unvarying value.

The consequence of this result is easily deduced. If projected on a plane perpendicular to the axis of the shell, the generating lines of the cone would become radii; and the groove which winds upwards from the base to the apex would be converted into a plane curve which, meeting those radii at an unvarying angle, would be neither more nor less than a logarithmic spiral. Conversely, the groove of the shell may be considered as the projection of this spiral on a conic surface.

Better still. Let us imagine a plane perpendicular to the axis of the shell and passing through its summit. Let us imagine, moreover, a thread wound along the spiral groove. Let us unroll the thread, holding it taut as we do so. Its extremity will not leave the plane and will describe a logarithmic spiral within it. It is, in a more complicated degree a variant of Bernouilli's "*Eadem mutata resurgo*": the logarithmic conic curve becomes a logarithmic plane curve.

A similar geometry is found in the other shells with elongated cones, Turritellae Spindle-shells, Cerithia, as well as in the shells with flattened cones, Trochidæ, Turbines. The spherical shells, those whirled into a volute, are no exception to this rule. All, down to the common Snail-shell, are constructed according to logarithmic laws. The famous spiral of the geometers is the general plan followed by the Mollusc rolling its stone sheath.

Where do these glairy creatures pick up this science? We are told that the Mollusc derives from the Worm. One day, the Worm, rendered frisky by the sun, emancipated itself, brandished its tail and twisted it into a corkscrew for sheer glee. There and then the plan of the future spiral shell was discovered.

This is what is taught quite seriously, in these days, as the very last word in scientific progress. It remains to be seen up to what point the explanation is acceptable. The Spider, for her part, will have none of it. Unrelated to the appendix-lacking, corkscrew-twirling Worm, she is nevertheless familiar with the logarithmic spiral. From the celebrated curve she obtains merely a sort of framework; but, elementary though this framework be, it clearly marks the ideal edifice. The Epeira works on the same principles as the Mollusc of the convoluted shell.

The Mollusc has years wherein to construct its spiral and it uses the utmost finish in the whirling process. The Epeira, to spread her net, has but an hour's sitting at the most, wherefore the speed at which she works compels her to rest content with a simpler production. She shortens the task by confining herself to a skeleton of the curve which the other describes to perfection.

The Epeira, therefore, is versed in the geometric secrets of the Ammonite and the *Nautilus pompilus;* she uses, in a simpler form, the logarithmic line dear to the Snail. What guides her? There is no appeal here to a wriggle of some kind, as in the case of the Worm that ambitiously aspires to become a Mollusc. The animal must needs carry within itself a virtual diagram of its spiral. Accident, however fruitful in surprises we may presume it to be, can never have taught it the higher geometry wherein our own intelligence at once goes astray, without a strict preliminary training.

Are we to recognize a mere effect of organic structure

in the Epeira's art? We readily think of the legs, which, endowed with a very varying power of extension, might serve as compasses. More or less bent, more or less out-stretched, they would mechanically determine the angle whereat the spiral shall intersect the radius; they would maintain the parallel of the chords in each sector.

Certain objections arise to affirm that, in this instance, the tool is not the sole regulator of the work. Were the arrangment of the thread determined by the length of the legs we should find the spiral volutes separated more wide-ly from one another in proportion to the greater length of implement in the spinstress. We see this in the Banded Epeira and the Silky Epeira. The first has longer limbs and spaces her cross-threads more liberally than does the sec-ond, whose legs are shorter.

But we must not rely too much on this rule, say others. The Angular Epeira, the Pale-tinted Epeira and the Diadem Epeira, or Cross Spider, all three more or less short-limbed, rival the Banded Epeira in the spacing of their lime-snares. The last two even dispose them with greater intervening distances.

We recognize in another respect that the organization of the animal does not imply an immutable type of work. Be-fore beginning the sticky spiral, the Epeiræ first spin an auxiliary intended to strengthen the stays. This spiral, formed of plain, non-glutinous thread, starts from the center and winds in rapidly-widening circles to the circumference. It is merely a temporary construction, whereof naught but the central part survives when the Spider has set its limy meshes. The second spiral, the essential part of the snare, proceeds, on the contrary, in serried coils from the circum-ference to the center and is composed entirely of viscous cross-threads.

Here we have, following one upon the other, by a sud-den alteration of the machine, two volutes of an entirely different order as regards direction, the number of whorls and the angle of intersection. Both of them are logarithmic spirals. I see no mechanism of the legs, be they long or short, that can account for this alteration.

Can it then be a premeditated design on the part of the Epeira? Can there be calculation, measurement of angles, gauging of the parallel by means of the eye or otherwise? I am inclined to think that there is none of all this, or at least nothing but an innate propensity, whose effects the animal is no more able to control than the flower is able to control the arrangement of its verticils. The Epeira practises higher geometry without knowing or caring. The thing works of itself and takes its impetus from an instinct imposed upon creation from the start.

The stone thrown by the hand returns to earth describing a certain curve; the dead leaf torn and wafted away by a breath of wind makes its journey from the tree to the ground with a similar curve. On neither the one side nor the other is there any action by the moving body to regulate the fall; nevertheless, the descent takes place according to a scientific trajectory, the "parabola," of which the section of a cone by a plane furnished the prototype to the geometer's speculations. A figure, which was at first but a tentative glimpse, becomes a reality by the fall of a pebble out of the vertical.

The same speculations take up the parabola once more, imagine it rolling on an indefinite straight line and ask what course does the focus of this curve follow. The answer comes: the focus of the parabola describes a "catenary," a line very simple in shape, but endowed with an algebraic symbol that has to resort to a kind of cabalistic number at variance with any sort of numeration, so much so that the unit refuses to express it, however much we subdivide the unit. It is called the number e. Its value is represented by the following series carried out *ad infinitum:*

$$e = 1 + \frac{1}{1} + \frac{1}{1\cdot2} + \frac{1}{1\cdot2\cdot3} + \frac{1}{1\cdot2\cdot3\cdot4} + \frac{1}{1\cdot2\cdot3\cdot4\cdot5} + \text{etc.}$$

If the reader had the patience to work out the few initial terms of this series, which has no limit, because the series of natural numerals itself has none, he would find:

$$e = 2.7182818 \ldots$$

With this weird number are we now stationed within the strictly defined realm of the imagination? Not at all: the catenary appears actually every time that weight and flexibility act in concert. The name is given to the curve formed by a chain suspended by two of its points which are not placed on a vertical line. It is the shape taken by a flexible cord when held at each end and relaxed; it is the line that governs the shape of a sail bellying in the wind; it is the curve of the nanny-goat's milk-bag when she returns from filling her trailing udder. And all this answers to the number e.

What a quantity of abstruse science for a bit of string! Let us not be surprised. A pellet of shot swinging at the end of a thread, a drop of dew trickling down a straw, a splash of water rippling under the kisses of the air, a mere trifle, after all, requires a titanic scaffolding when we wish to examine it with the eye of calculation. We need the club of Hercules to crush a fly.

Our methods of mathematical investigation are certainly ingenious; we cannot too much admire the mighty brains that have invented them; but how slow and laborious they appear when compared with the smallest actualities! Will it never be given to us to probe reality in a simpler fashion? Will our intelligence be able one day to dispense with the heavy arsenal of formulæ? Why not?

Here we have the abracadabric number e reappearing, inscribed on a Spider's thread. Let us examine, on a misty morning, the meshwork that has been constructed during the night. Owing to their hygrometrical nature, the sticky threads are laden with tiny drops, and, bending under the burden, have become so many catenaries, so many chaplets of limpid gems, graceful chaplets arranged in exquisite order and following the curve of a swing. If the sun pierce the mist, the whole lights up with iridescent fires and becomes a resplendent cluster of diamonds. The number e is in its glory.

Geometry, that is to say, the science of harmony in space, presides over everything. We find it in the arrangement of the scales of a fir-cone, as in the arrangement of an

Epeira's lime-snare; we find it in the spiral of a Snail-shell, in the chaplet of a Spider's thread, as in the orbit of a planet; it is everywhere, as perfect in the world of atoms as in the world of immensities.

And this universal geometry tells us of an Universal Geometrican, whose divine compass has measured all things. I prefer that, as an explanation of the logarithmic curve of the Ammonite and the Epeira, to the Worm screwing up the tip of its tail. It may not perhaps be in accordance with latter-day teaching, but it takes a loftier flight.

Like Fabre, E. E. Slosson was a fascinated observer of the oddities of natural science. He was the director of Science Service and the author of Creative Chemistry, *a highly popular book of the past generation. The charm and clarity of his writing are exemplified in this little essay on "Why Kitty Lands Butter-Side Up."*

WHY KITTY LANDS BUTTER-SIDE UP

E. E. SLOSSON

THE SLOW motion-picture camera has enabled the scientist to solve the age-old mystery as to how a freely falling cat always manages to land on its feet even when dropped from a comparatively small height.

They have found that puss uses the formula $a = \dfrac{L \times L}{(k) \, \epsilon \, (m \, r^2)}$, which is "the angular acceleration of a rigid body under the action of a resultant torque."

The mathematics involved is rather complicated for the layman and would probably annoy the cat somewhat also if she paused to figure it out on the way down. The whole problem, and incidentally the cat, turns on the principle that it takes more power to rotate an object through a large circle than a small one.

If one were to imagine a body—Figure IV-3—consisting of a rod CD, to the ends of which are hinged four weights, A and A′ and B and B′, and if by means of some machinery inside the rod it could be made to twist, the two sets of weights would quite evidently rotate in opposite directions. Weights A and A′, being farther from the axis of rotation,

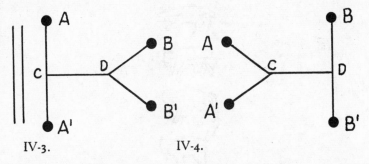

IV-3.   IV-4.

would have more leverage and would remain almost stationary, whereas B and B′ would take new positions. If A and A′ are now brought near together and B and B′ swung apart with the same twist applied, the latter two hold their positions, but the first two weights turn—Figure IV-4.

The motion picture revealed that the first part of Kitty's technic is simultaneously to extend the hindlegs and tail perpendicular to the axis of her body and to draw the forelegs close in. A twisting strain is now applied through the body and results in the closely held forequarters rotating nearly ninety degrees in advance of the hindquarters. Then, by drawing in the hindlegs and tail, extending the forelegs and exerting another torsional stress in a direction opposite to the previous one, the hindquarters are brought around and the cat is ready to land on her feet, without using any mathematics at all.

IV-5.

The desires of gamblers to get more exact information about
the turns of cards and the rolls of dice impelled mathematicians of the Middle Ages to found the branch of
mathematics known as probability theory. From these simple
beginnings it has grown to enormous proportions and has
become the foundation on which a large part of modern
industrial and business activity rests. Insurance in all its forms
is an obvious example of this dependence. Production control and communications systems are others. And statistical
methods have become among the most powerful tools of
the modern experimental worker in both the biological and the

physical sciences. M. J. Moroney, a graduate of the University of London, a statistician specializing in industrial problems, and a Senior Lecturer at the Leicester College of Technology and Commerce, here describes the fundamentals of his craft. The author points out that there may be truth in the statement that "there are lies, damned lies, and statistics," but only for those who do not know how to interpret the figures correctly.

THE LAWS OF CHANCE

M. J. MORONEY

> "Quoth she: 'I've heard old cunning stagers
> Say fools for arguments use wagers.'"
> —s. BUTLER (*Hudibras*)

THERE ARE certain notions which it is impossible to define adequately. Such notions are found to be those based on universal experience of nature. Probability is such a notion. The dictionary tells me that "probable" means "likely." Further reference gives the not very helpful information that "likely" means "probable." It is not always that we are so quickly made aware of circularity in our definitions. We might have had an extra step in our circle by bringing in the word "chance," but, to judge from the heated arguments of philosophers, no extension of vocabulary or ingenuity in definition ever seems to clear away all the difficulties attached to this perfectly common notion of probability.

In this chapter we shall try to get some idea of what the statistician has in mind when he speaks of probability. His ideas are at bottom those of common sense, but he has them a little more carefully sorted out so that he can make numerical statements about his problems instead of vague general comments. It is always useful when we can measure things on a ruler instead of simply calling them "big" or "small."

The Probability Scale

We measure probability by providing ourselves with a scale marked zero at one end and unity at the other. (In reading what follows, the reader will do well to keep Figure IV-6 constantly before his attention.) The top end of the scale, marked unity or 1, represents absolute certainty. Any proposition about which there is absolutely no doubt at all would find its place at this point on the scale. For example: The probability that I shall one day die is equal to unity, because it is absolutely certain that I shall die some day.*

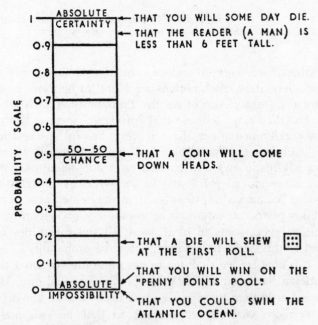

IV-6.   *The Probability Scale*

*Quia pulvis es, et in pulverem reverteris (Gen. iii, 19).

The mathematician would here write $p = 1$, the letter p standing for probability. The bottom end of the scale, marked zero or 0, represents absolute impossibility. For example: The probability that I should succeed in an attempt to swim the Atlantic is zero, because failure would be absolutely certain. The statistician would here write $p = 0$.

If all the affairs of life were as clear-cut as this, statisticians would be out of a job, and scientific research would shoot ahead at an intolerable rate, losing most of its interest. Life and nature may be simple enough to the Almighty who designed them and keeps them going, but to the human mind there is presented an unending stream of problems that cannot be given a clear-cut answer of the type $p = 1$ or $p = 0$. The doctor knows that penicillin is excellent for your particular disease, but he cannot absolutely guarantee that you will be cured by using it. At most he can be very sure. He may say that for all practical purposes he is prepared to put $p = 1$ for your recovery. But this is an approximation; we have already slipped from the realm of absolute certainty. In fact, we may suppose, $p = 0.999$. What the doctor then says is: "We may without noticeable error put $p = 1$." Figure IV-6 shows the sort of position occupied on the scale of probability by various common affairs. The thing to notice is that there is no greater certainty than $p = 1$, and nothing less likely than $p = 0$.

So far, then, we have set up our scale on which the probability of events may be specified. How do we arrive at an actual measure of the probability of any real life event? There are two main ways, and we shall consider them in turn.

A Priori Probabilities

These are probabilities which we feel certain we can specify in magnitude from consideration of the very nature of the event. For example: The probability that if I spin a penny it will come down heads is easily and sensibly guessed to be $p = \frac{1}{2}$. Intuitively, we feel that the probability of heads comes exactly halfway along the scale in Figure IV-6. We may look at it from another common-sense point of view. There are two ways in which the spin may turn up: head or tail. Both these ways are equally likely. Now it is absolutely certain that the coin will finish up head *or* tail, i.e., for head *or* tail $p = 1$. The total probability $p = 1$ may be shared between the two possible results equally, giving $p = \frac{1}{2}$ for a head, and $p = \frac{1}{2}$ for a tail.

In like manner, there are six equally likely results if we roll an unbiased die. Certainly the result is bound to be one of the six equally probable results. The probability of getting *some* number is $p = 1$. Dividing this total probability between the six possibilities, we say that there is a probability of $p = \frac{1}{6}$ for each of the possible results. (We ignore in all cases the preposterous suggestion that the coin will land on its edge or the die stand up on one corner.)

Empirical Probability

The problem of probabilities in card and dice games may be tackled from another point of view. Say, having made a die, we roll it 600 times. We should *expect* that each face would have shown uppermost 100 times. What do we mean by "expect"? We don't really expect anything of the sort. In fact, we should be rather surprised at the "coincidence" if any practical trial gave a result in perfect agreement with our "expectation." What we really expect is that each face would turn up *roughly* 100 times—not too roughly, of course, or we should suspect bias; nor too exactly, either, or we might suspect jiggery-pokery. This suggests to us another way of measuring the probability of an event: by counting the number of times the event occurs in a certain number of trials. We take it that a very long series will give a closer indication of the probability than a short series. We believe from our experience of things that while short trials are easily upset by "chance" a long trial is protected by the mysterious laws of this very same "chance." We may express the empirical probability of an event as:

$$\text{Probability} = \frac{\text{Total number of occurrences of the event}}{\text{Total number of trials}}$$

Thus, for example, if a surgeon performs a certain operation on 200 people and 16 of them die, he may assume the probability of death to be $p = \frac{16}{200} = 0.08$. This empirical method of finding probabilities as the ratio of the number of occurrences to the total number of trials is the method that has to be used in many fields of research.

Having seen how probabilities may be measured, we must now consider some of the laws of probability, so that we can analyse more complex situations.

Addition Law

Consider the phrase "Heads I win; tails you lose." This is the simplest possible illustration of the Law of Addition. To calculate my total chance of winning, I have, according to this law, to add up the probabilities of each of the several ways in which I may win. In the first place, I shall win if the coin turns up heads, and this has $p = \frac{1}{2}$. In the second place I shall also win if the coin turns up tails, and this also has $p = \frac{1}{2}$. Adding the two probabilities together, we see that the total probability of my winning is $p = \frac{1}{2} + \frac{1}{2} = 1$. That is, it is absolutely certain that I shall win.

The probability that an event will occur in one of several possible ways is calculated as the sum of the probabilities of the occurrence of the several different possible ways.

It is assumed that the occurrence of the event in one way excludes the possibility of its occurrence in any of the other possible ways, on the occasion in question.

As a simple example, let us suppose that 10 Englishmen, 8 Irishmen, 2 Scotsmen, and 5 Welshmen apply for a job to which only one man will be appointed. Altogether there are 25 applicants. Let us suppose that the interviewing board are completely unable to agree with each other on the respective merits of the applicants, and so decide to draw a name out of the hat. The probability of the job going to an Englishman will evidently be $\frac{10}{25}$; to a Scotsman, $\frac{2}{25}$; to a Welshman, $\frac{5}{25}$; and to an Irishman, $\frac{8}{25}$. Then the Law of Addition gives us the following results:

Probability of a Celt $= \frac{2}{25} + \frac{5}{25} + \frac{8}{25} = \frac{15}{25} = 0.6$.

Probability of native of Great Britain

$$= \frac{10}{25} + \frac{2}{25} + \frac{5}{25} = \frac{17}{25} = 0.68$$

Probability of NOT a native of Great Britain $= \frac{8}{25} = 0.32$

Multiplication Law

We shall now prove, to the no little satisfaction of the fair sex, that every woman is a woman in a billion. It is hoped that menfolk will find salve for their consciences in this scientific proof of the age-old compliment. ("Statistics show, my dear, that you are one in a billion.") It will be obvious to the reader that the more exacting we are in our demands, the less likely we are to get them satisfied. Consider the case of a man who demands the simultaneous occurrence of many virtues of an unrelated nature in his young lady. Let us suppose that he insists on a Grecian nose, platinum-blonde hair, eyes of odd colours, one blue and one brown, and, finally, a first-class knowledge of statistics. What is the probability that the first lady he meets in the street will put ideas of marriage into his head? To answer the question we must know the probabilities for the several different demands. We shall suppose them to be known as follows:

Probability of lady with Grecian nose: 0.01

Probability of lady with platinum-blonde hair: 0.01

Probability of lady with odd eyes: 0.001

Probability of lady with first-class knowledge of statistics: 0.00001

In order to calculate the probability that all these desirable attributes will be found in one person, we use the Multiplication Law. Multiplying together the several probabilities, we find for our result that the probability of the first young lady he meets, or indeed any lady chosen at random, coming up to his requirements is $p = 0.000\,000\,000\,001$, or precisely one in an English billion. The point is that every individual is unique when he is carefully compared, point by point, with his fellows.*

We have considered here the case of the simultaneous occurrence of events. The Multiplication Law is also used when we consider the probability of the occurrence of two or more events in succession, even where the successive events are dependent. Consider the following example: A bag contains eight billiard balls, five being red and three white. If a man selects two balls at random from the bag, what is the probability that he will get one ball of each color? The problem is solved as follows:

The first ball chosen will be either red or white, and we have:

Probability that first ball is red = $\frac{5}{8}$. If this happens, then there will be four red balls and three white balls in the bag for the second choice.

Hence the probability of choosing a white after choosing a red will be $\frac{3}{7}$.

The Multiplication Law tells us that the probability of choosing white after red is $\frac{5}{8} \times \frac{3}{7} = \frac{15}{56}$.

In like manner, the probability of the first ball out being white is $\frac{3}{8}$.

This will leave two white balls in the bag for the second choice.

Hence the probability of choosing a red ball after choosing a white one will be, by the Multiplication Law: $\frac{3}{8} \times \frac{5}{7} = \frac{15}{56}$.

Now the man will have succeeded in getting one ball of each color in either case. Applying the Addition Law,

*The different applications of the Laws of Addition and Multiplication of probabilities may be remembered in terms of betting on horse racing. If I bet on two horses in the same race the probability of my winning is the *sum* of the probabilities for winning on each of the two horses separately. If I have an "accumulator bet," i.e., bet on one horse in the first race and direct that my winnings, if any, be placed on one horse in the second race, then my chance of winning the accumulator bet is the *product* of the probabilities that each of my chosen horses will win its own race.

we find the probability of his success to be
$$\tfrac{15}{56} + \tfrac{15}{56} = \tfrac{30}{56} = \tfrac{15}{28} = 0.535.$$

The Addition Law and the Multiplication Law are fundamental in Statistics. They are simple; but sufficient to carry us a long way, if we make good use of them.

What we have discussed so far is known as the Direct Theory of probability. Basically, all the problems commonly met with in this branch of the subject turn on counting the number of ways in which events can occur. For example: if we ask ourselves what is the probability that three pennies on being tossed will all show heads, we can arrange all the possible results in a table as follows:

RESULT	1ST COIN	2ND COIN	3RD COIN
3 Heads	H	H	H
2 Heads	H	H	T
	H	T	H
	T	H	H
2 Tails	T	T	H
	T	H	T
	H	T	T
3 Tails	T	T	T

In the table, H represents head and T represents tail. If we assume all the possible results to be equally likely, then of the eight possible results, only one will be a success. Hence the probability of all three coins showing heads is $p = \tfrac{1}{8}$. In like manner, the probability is again $p = \tfrac{1}{8}$ that all the coins will show a tail. Hence, by the Addition Law, the probability of three heads *or* three tails will be $p = \tfrac{1}{8} + \tfrac{1}{8} = \tfrac{1}{4}$.

This is a suitable point to introduce some fallacious arguments for the reader to consider.

Fallacious argument Number 1. There are two possible results: either all the coins show alike or they don't. Hence the probability of all the coins showing the same face is $p = \frac{1}{2}$.

Fallacious argument Number 2. There are four possible results: all heads, all tails, two heads and a tail, or two tails and a head. Two of these results would be satisfactory. Hence the probability of all the coins showing the same face will be $p = \frac{1}{2}$.

These arguments are invalid because they assume events to be equiprobable which in fact are not so. Inspection of the table will show that there is only one way of getting the result three heads. There is similarly only one way of getting the result three tails. But the result two heads and a tail can occur in three different coin arrangments, as also can the result two tails and a head.

It is a simple enough matter to write out all the possible arrangments where these are relatively few in number. The introduction of permutations in football pools recognized the difficulty of writing out complex cases by the punter and the enormous labor of checking them. It will be useful to spend a few moments on the idea of Permutations and Combinations.

Combinations and Permutations

Suppose a race were run by seven children and that we attempted to predict the first three children home. It is one thing to name the three children irrespective of their placing, and quite another to get not only the first three correct but also their placing. When a problem concerns groups without any reference to order within the group it is a problem in combinations. When the problem asks us to take arrangements into account it is a problem in permutations. Thus what is commonly called a combination lock is really a permutation lock, since order is vitally important. On the other hand, the football pools fan who enters six teams for the "four aways" and writes on his coupon "Perm. 4 from 6, making 15 lines at 6d. Stake 7s. 6d.," is really talking about a combination, since there is no question of arranging the correct four teams in any way. It is sufficient to name them in any order whatsoever. The "penny points pool," on the other hand, is indeed a permutation; it is not sufficient to get the correct number of wins away and at home and the correct number of draws; correct arrangement within the column is essential.

Permutations are more numerous than combinations, for each combination can be permuted. As an example the group of letters *ABC* which make a single combination, whatever their order, gives rise to six permutations, viz. *ABC, ACB, BCA, BAC, CAB, CBA*.

We shall now give some of the main results in the theory of permutations and combinations with simple illustrations of each type.

Simple Cases of Choices

If there are m ways of performing one operation, n ways of performing a second operation, and p ways of performing a third operation, then there are $N = m \times n \times p$ ways of performing the whole group of operations.

> *Example.* A man travelling from Dover to Calais and back has the choice of ten boats. In how many ways can he make the double journey, using a different boat in each direction?

Going, he has the choice of all ten boats, i.e., the first operation (going) can be performed in $m = 10$ ways. Coming back, he will only have nine boats to choose from, i.e., the second operation (returning) can be performed in $n = 9$ ways. Hence, there are $N = m \times n = 10 \times 9 = 90$ ways of making the double journey.

> *Example.* How many lines would be required for a full permutation on a fourteen-match "penny points pool"?

Regarding the forecasting of each match as an operation, we have fourteen operations to perform. Each operation can be dealt with in three ways, viz. 1, 2, or X. Hence the total number of ways of forecasting the result will be

$$N = 3 \times 3 \times 3 \times 3 \times 3 \times 3 \times 3 \times 3 \times$$
$$3 \times 3 \times 3 \times 3 \times 3 \times 3 = 4,782,969.$$

This number of entries at 1d. per line would cost roughly £20,000. It is the author's considered opinion that the amount of skill one can bring to bear in forecasting is a relatively negligible quantity. In so far as this is true, no amount of permuting is likely to be of great assistance while the old lady with a pin is in the running. It would be salutary for readers of expert permutationists in the news-papers to remember that armies of gullible fans, sending in massive permutations week after week, are bound to pro-duce some successes for the expert to advertise. The real test is: how many weeks in the season does the columnist himself bring home a really substantial prize?

Example. A factory call-light system has four colors. The lights may be on one, two, three, or four at a time. If each signal combination can be made to serve for two people, being steady for one and flickering for the other, how many people can be accommodated on the system?

This problem is very easily dealt with. Ignore for the moment the question of flickering. There are two ways of dealing with the first lamp—switch it on or leave it off. There is the same choice for each lamp. Evidently, then, the total number of ways in which the system may be set will be $N = 2 \times 2 \times 2 \times 2 = 16$. But this would include the case where all the lights were left off. We must leave this case out as being of no use as a signal. We are left with fifteen signals. Each of these fifteen signals may be either steady or flickering, so the system can accommodate thirty people.

Permutations

If all the things to be arranged are different, it is very simple to calculate the total number of permutations.

Example. In how many ways can the letters of the word BREAD be arranged?

In the first position we can have a choice of five letters. Having filled the first place, we shall be left with a choice of four letters for the second place. In turn, there will be a choice of three letters for the third place, two letters for the fourth place, and, finally, only one letter to go into the last place. Applying our previous rule we find the total number of ways of arranging the letters is $N = 5 \times 4 \times 3 \times 2 \times 1 = 120$.

Example. How many three-letter words can be made using the letters of the word BREAD?
Similar reasoning to that used above yields the answer

$$N = 5 \times 4 \times 3 = 60$$

The mathematician has a simple piece of shorthand for permutations. In our first example we were arranging five things in every possible way, each thing appearing in each arrangement, i.e., we were arranging, or permuting, five things in groups of five at a time. The shorthand for this is 5P5. In the second problem we were arranging the five things in groups of three. The shorthand for this is 5P3. The letter P stands for "the number of permutations." The number before the P tells us how many things we have to choose from; and the number after the P tells us how many things are to be in each arrangement. Thus if we saw 43P7, we should know that there were forty-three things to be made up into every possible arrangement (order counting), there being seven things in each arrangement.

It is convenient here to introduce one other piece of shorthand, which is easy to understand and which saves a great deal of time in writing things down. It will be remembered that the result for our first problem in permutations (arranging the letters of the word BREAD in every possible five-letter arrangement) was $N = 5 \times 4 \times 3 \times 2 \times 1$. Here we have multiplied together a string of numbers, starting with 5, each number being one less than the one before it, the last number in the sequence being 1. Such an arrangement is called a "factorial." One or two examples will make the meaning clear.

Factorial $5 = 5 \times 4 \times 3 \times 2 \times 1 = 120$

Factorial $7 = 7 \times 6 \times 5 \times 4 \times 3 \times 2 \times 1 = 5,040$, and so on. The shorthand sign for the factorial of a number is made by writing an exclamation mark after the number. Thus factorial 7 is written 7! and factorial 93 is written 93! The use of this factorial sign will enable us to write down further results in the theory of permutations and combinations compactly.

What happens if we have to make permutations of things that are not all different? Obviously we shall have to allow for the fact that the identical things can be interchanged without disturbing the permutation.

If we have n things, p being alike of one kind, q alike of another kind, and r alike of another kind still, then the total number of ways in which all the n things can be arranged so that no arrangement is repeated is:

$$N = \frac{n!}{p! \times q! \times r!}$$

Example. How many different permutations may be made each containing the ten letters of the word STATISTICS? Here we have the letter S three times, the letter T three times, the letter I twice, and the letters A and C once each. Applying our rule, we get:

$$N = \frac{10!}{3! \times 3! \times 2!} = \frac{10.9.8.7.6.5.4.3.2.1}{3.2.1 \times 3.2.1 \times 2.1} = 50,400$$

Combinations

It remains for us now to consider the problem of calculating the number of combinations (i.e., irrespective of order) which can be made from a group of things. We have already seen that any combination can give rise to a set of permutations, the combination ABC yielding, for example, the six permutations ABC, ACB, BCA, BAC, CAB, CBA. Very little thought is required to see that a combination of n things can generate n! permutations. Thus in any problem, if we knew the number of combinations that could be made, and knew the number of permutations to which each combination could give rise, we should know that the total number of permutations was equal to the number of combinations multiplied by the number of permutations within a combination. *Number of Combinations × Number of permutations within a Combination = Total Number of Permutations.*

Just as, previously, we denoted the number of permutations of five things taken three at a time by the symbol 5P3, so now we shall denote the number of combinations of five things taken three at a time by the shorthand symbol 5C3. The letter C stands for "the number of combinations that can be made." The number before the C tells us how many things we have to choose from, and the number after the C tells us how many things are to appear in each combination. The number of combinations of n things taken r at a time will thus be denoted by nCr and the number of permutations of n things taken r at a time will be denoted by nPr. Now we know that r things forming a combination can give us r! permutations, so we have our previous result in mathematical form as:

$$nCr \times r! = nPr$$

from which, dividing both sides by r!, we find that the number of combinations of r things at a time chosen from a group of n things is to be calculated as:

$$nCr = \frac{nPr}{r!}$$

It is clear, too, that whenever we make a choice of items to include in a combination, we thereby automatically also make a choice of the remaining items to exclude from our combination. For example, if we are forming combinations of three things from five things, every time we choose a group of three (to include) we also choose a group of two, the remainder (to exclude). It follows that

$$nCr = nC(n - r)$$

This result is often useful in calculating, as a time saver.

Example. From a group of seven men and four ladies a committee is to be formed. If there are to be six people on the committee, in how many ways can the committee be composed (a) if there are to be exactly two ladies serving, (b) if there are to be at least two ladies serving?

Consider first the case where there are to be *exactly* two ladies. There are two distinct operations to be performed: (i) choosing the ladies, (ii) choosing the men. The number of ways of choosing two ladies from four ladies is $4C2 = \dfrac{4P2}{2!} = \dfrac{4 \times 3}{2 \times 1} = 6$. The number of ways of choosing four men to make the committee up to six is $7C4 = \dfrac{7P4}{4!} = \dfrac{7 \times 6 \times 5 \times 4}{4 \times 3 \times 2 \times 1} = 35$. Hence there are six ways of performing the first operation (choosing ladies) and thirty-five ways of performing the second operation (choosing men). The total number of ways of selecting the committee is therefore $N = 6 \times 35 = 210$.

Consider, now, the second problem, where there are to be at least two ladies. In addition to the 210 ways of having exactly two ladies, we shall have the number of ways in which we can have three ladies and three men, or four ladies and two men (there are only four ladies available).

Arguing exactly as before, we find the number of ways of having three ladies and three men is $4C3 \times 7C3 = 140$, and the number of ways of having four ladies and two men is $4C4 \times 7C2 = 21$. Adding up all these results, we find that the total number of ways of having at least two ladies on the committee is $210 + 140 + 21 = 371$.

Permutations and combinations make nice brain-teasers. The arithmetic is dead easy, but it is essential to think very clearly.

One of the bugaboos of modern civilization is the belief that science is creating Frankensteins that will perform many of the functions of man and perhaps in the end destroy him. One such Frankenstein is the electronic computer— labeled "thinking machine" or "electronic brain" in popular parlance. In a foreword to Minds and Machines, from which "Computing and Other Machines" has been selected, Professor C. A. Mace dismisses such dangers. He does admit, however, that some of the new machines are "rather frightening," and that no one can rule out the possibility that the world's championship in chess will one day be held by a machine, or that a machine will in the future be able to compose an acceptable sonnet.

Even today, analogue and digital computers are performing miracles of computation, answering mathematical questions in hours or days which would require decades of work by whole staffs of human calculators. These machines have not yet been able to pose—and prove or disprove—new mathematical theorems, but even this feat is not an impossibility.

Wladyslaw Sluckin was born at Warsaw in 1919 and received his early education there. He obtained his Ph.D. from the University of London in 1955. He is a qualified engineer, but his present interests lie in experimental and comparative psychology. He finds points of similarity between the workings of the human and of the electronic brain. He here introduces us to the various kinds of computers which are currently being actively developed.

COMPUTING AND OTHER MACHINES

WLADYSLAW SLUCKIN

MATHEMATICAL MACHINES are classified into two categories. To the one belong the so-called analogy (or analogue) machines which are basically mathematical instruments, exemplified in a simple form by the slide-rule; to the other belong digital computers, a simple representative of which is the familiar desk calculating machine.

We shall concern ourselves with digital calculating machines rather more than with the other category for two reasons. Firstly, it is machines of this class which are most frequently referred to as electronic brains; machines of this class are said, sometimes in jest and sometimes seriously and not without some justification, to be capable of thinking. Secondly, there are some important and suggestive similarities between the working of digital computers and the working of nervous systems in living creatures.

However, some forms of analogy machines, too, have been or may be called "thinking machines." Analogy machines are also of interest here because of certain affinities between what are known as "continuous operations" characteristic of these machines and certain features of the functioning of organisms.

Mechanical Counters

The elementary basis of operation of any digital computer, however complex, is the same as that of the old-fashioned abacus. In the abacus any number smaller than 10 is represented by so many beads. For larger numbers each digit is represented by a set of beads, the number of beads in each set corresponding to the value of the digit. Because we use

297

the decimal system for number notation, our abaci work to the base of 10. It is theoretically possible to have, say, a duodecimal abacus working to the base of 12, or a binary abacus working to the base of 2, and so on. The method of number notation and, therefore, of abacus arrangement is a matter of tradition, convention, and convenience.

In an abacus the counting of beads is the basis of addition and subtraction. In digital computers all mathematical calculations depend ultimately on counting. Instead of beads the objects of counting may be cogs of a gear wheel, or events such as electrical pulses. But counting is the basic operation. By contrast, there is no counting of discrete quantities in analogy machines.

Consider an arrangement for mechanically counting the number of miles covered by a vehicle on a journey. It is simple enough to arrange that each time a wheel turns a full circle, some sort of a signal is registered. The total number of signals registered is proportional to the mileage covered. To compute the mileage, we need only know the length of the circumference of the wheel. For our convenience we may calibrate the registering meter directly in miles or fractions of a mile.

Even the simplest counter performs what is sometimes called the operation of "carrying." A domestic electric meter counts the complete revolutions of a small aluminium disk, the speed of which is proportional to the rate of electric

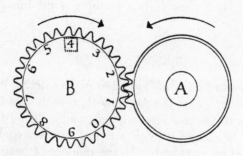

IV-7. *Counting by "Carrying"*

power consumption. While the disk itself is an integrator and an analogy mechanism, the counting is digital. In the simplest case such a meter would register "1" after one complete revolution of the disk; it would register "2" after two revolutions, and so on. This may be done in the manner shown in Figure IV-7.

Gear wheel B has teeth right round its periphery. Wheel A is of equal size but is stripped of all but one-tenth of its teeth. Wheel A is permanently fixed to the shaft of the revolving disk. Suppose it turns anti-clockwise, as shown. Each time wheel A makes a complete revolution, it causes wheel B to move in the clockwise direction a distance equal to the space separating two consecutive numbers upon it. Imagine that wheel B is covered up, and that there is a little window in the cover where a dotted square is drawn. Then, to start with, "0" is seen through the window. After wheel A has been once round "1" appears, two revolutions of wheel A brings "2" to the window, and so on. Our simple arrangement can only count as far as 9.

Of course, another wheel, exactly similar to A, can be attached to the shaft of wheel B. Though this is not shown in Fig. IV-7, we may refer to such a wheel by some other letter, say C. This one-tenth stripped wheel can be geared with another wheel—call it D—exactly similar to B. Then, when wheel A has been ten times round, the window upon B will show "0"; but since wheel B has now been once right round, a similar window upon D will read "1." Disk D will show a reading of "2" only after A has been right round twenty times, and so on. We see, then, that B reads units and D reads tens of revolutions. Hundreds, thousands, etc., may be arranged to be read in a similar manner.

Digital Adders and Multipliers

Addition of two numbers consists of first counting one number and then the other. Basically, an adder consists of two counters. Before adding begins both counters are set at zero. The first of the two numbers to be added up is then registered on the first counter. To add the second number to the first, both counters must now be locked and used simultaneously. They are stopped when the second counter has registered the second quantity. By then the first counter will have counted both the numbers, thus having added them together. In this manner it will have recorded the sum of the two numbers.

The arrangement described above is achieved mechanically by means of a simple ratchet. The first counter moves freely alone; but the second counter drags with it the first. Devices such as magnetic clutch counters, punched cards and others can be used for the purpose of effecting addition. There are also relay and electronic adders. All these add up numbers by counting first one and then the next. The basic principle of digital addition has been known since Pascal first expounded it in the seventeenth century.

Shortly afterwards, the philosopher and mathematician Leibnitz showed how a mechanical multiplier could be made. Multiplication is done like addition with the help of yet another counter. 3×5 means 3 added 5 times, or 5 added 3 times. Therefore what is required for multiplication is an adder equipped with a means of counting the number of additions. Thus 3 added twice means $3 + 3$; here only two counters are needed. But when 3 is added 5 times or any other number of times an extra counter is required to inform the operator when to stop adding, or preferably to control the adding process by stopping it automatically at the appropriate moment.

Subtracting consists of adding the second quantity in reverse. Division is the reverse of multiplication, though in practice division mechanisms may be separate from multipliers. Thus all the four fundamental operations of arithmetic are based on counting.

Arithmetical Operations by Electrical Methods

The operations of arithmetic performed digitally by geared cogs may also be carried out by electrical means. Either electro-magnetic (electro-mechanical) relays or thermionic valves may be used.

The principle of a relay is shown in Figure IV-8. The electro-magnetic relay is a switching device: when no current flows through the coil, the iron core remains unmagnetized and the spring holds the moving-iron armature back. The circuit to be controlled is then continuous between

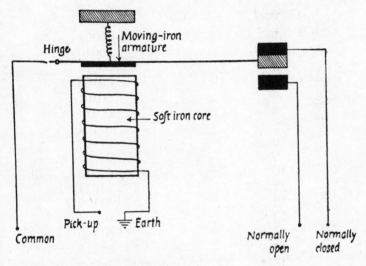

IV-8.   *Electro-Magnetic Relay*

the "common" terminal and the one marked "normally closed." When current does flow in the coil, the iron core is magnetized. The electro-magnet, then, pulls the moving-iron, and as a result the switch is thrown over from "normally closed" position to "normally open."

Thus, a circuit controlled by a relay is in one of the two positions; these may be denoted by "1" and "0." Several relays side by side, some in position "1" and some in position "0," may represent a number. A number so represented is expressed in what is called a binary system of notation, and not in the familiar decimal one.

There are a number of possible ways of representing

numbers in the binary system. Consider one way. It is as
follows. Number 1 in the decimal system is made to correspond to 1 in the binary, 2 in the decimal to 10 in the
binary, 3 (decimal) = 11 (binary), 4 (decimal) = 100
(binary), 5 (decimal) = 101 (binary), etc. For a binary
notation of this type, numbers from 1 to 16 are set out
below.

Decimal	Binary
1	00001
2	00010
3	00011
4	00100
5	00101
6	00110
7	00111
8	01000
9	01001
10	01010
11	01011
12	01100
13	01101
14	01110
15	01111
16	10000

Remembering that *1* and *0* denote relay positions, it
will be seen that if a set of relays is arranged thus: 00011,
it represents the number 3. If, then, a signal 00100 (i.e., 4)
is sent in, the relays will be arranged thus: 00111 (i.e., 7).
We see that

$$+\frac{00011 \quad (3)}{00100 \quad (4)}$$
$$\overline{00111 \quad (7)}$$

Now start, say, with a relay arrangement of 01001 (9)
and add to it 00101 (5). The relays are so interconnected
that when "1" signal reaches a relay in *1* position, the

result on that relay is *0*, but the relay immediately in front receives signal "1." Thus

$$+\begin{array}{l}01001 \quad (9)\\ 00101 \quad (5)\end{array}$$
$$\overline{01110 \quad (14)}$$

If the relay in front is already in *1* position, then on receiving a "1" signal it becomes *0*, and a "1" signal is sent forward to the next relay in front, and so on.

Such a dyadic—as it is called—counting and adding system may be used as a basis of all other digital operations; thus, just as well as a mechanical counter, an electro-magnetic relay may be employed as a fundamental unit of digital calculating machines.

A thermionic valve such as the triode shown diagrammatically in Figure IV-9 can often be used with advantage as a basic element of a digital computer. Provided that an adequate voltage is kept up, electrons will flow from the cathode to the anode and right round the circuit. But a sufficiently negative grid bias voltage can prevent the electrons from reaching the anode.

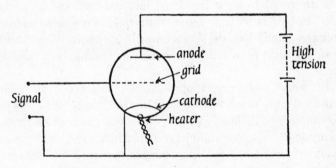

IV-9.   *A Triode Circuit*

Thus, a suitable signal arriving at the grid can break the main circuit current. Depending, therefore, on the value of the grid voltage, a current does or does not flow in the main circuit. In this way a triode may be used as a switching or trigger device much like a relay. The speed of a change-over in the relay is about one-hundredth of a second, while the speed of change-over in the valve is only about one-millionth of a second; and a simple valve may be cheaper than a relay.

Punched-Card Machines

Towards the end of the last century Hollerith in America developed a mechanical sorter. Its original use was in connection with the American Census. In such devices numerical information relating to large numbers of people or things is coded in the form of patterns of holes in cards, each unit about which information is available being assigned a separate card. All punched-card machines have this much in common. Different makes use different methods of taking cognizance of the information that is available on the cards in the form of hole arrangements. Various common features of the cards, that is various criteria for sorting may be applied to a single pack of punched cards: first one, then another, and so on, until as much general information as necessary is abstracted from the cards.

Patterns of holes in modern punched-card machines are used not merely as a basis of sorting. They also actuate counters; and, like desk machines, punched-card machines perform all the usual operations of adding, multiplying, etc.

In some commercial applications, however, the older punched-card machines appear to be already out of date. The most recent development is the adaptation of electronic computers to accountancy and production control. Such machines can be used, for example, to work out the pay-roll of a large firm, calculating and deducting for each individual P.A.Y.E., National Insurance and other items, all on the scales appropriate to the individual. A recently developed punched-card machine is used by the Austin Motor Company to control the operations of final vehicle assembly. Punched cards carrying appropriate instructions not only operate the feed of sub-assemblies on to the final line, but also the preparation of the necessary documents for each vehicle.

Automatic Digital Computers

A striking development in the field of digital machines took place when large computers began to be constructed early in the nineteen-forties. The elementary functioning of these machines did not depart from that of the desk calculators in that numbers were operated upon directly; that is, counting, effected by whatever means, was and is the basic operation. The decisive progress, when it occurred, lay in the fact that the new machines could perform longer sequences of simple operations, and that these sequences were adjustable to a high degree, and could be determined as necessary by the requirements of the problem in hand.

The conception of such an automatic, sequence-controlled digital computer is not new. Over a hundred years ago Babbage, a professor of Mathematics at Cambridge, conceived such a machine; he named it the "Analytical Engine." The idea of this machine has often been called in the literature "Babbage's dream," for Babbage's design was never put into practice. But even the modern punched-card machine may be said to be up to a point the realization of Babbage's dream.

Over one hundred years had passed since Babbage expounded his scheme before a machine very much the same as Babbage's in conception, though quite different in construction, was made. The designer was Professor Aiken of Harvard, and the makers the International Business Machines Corporation. The computer was named the Automatic Sequence-Controlled Calculator, Harvard Mark I. The calculating elements of this machine consist of mechanical counters which are driven through electro-magnetic clutches. The clutches are controlled by electro-mechanical relay circuits. Harvard Mark I machine has been described more than once as a "mechanical brain."

Information goes into this machine by means of dial-switches, punched cards, or tape. Information comes out as punched cards, or tape, or electrically typewritten paper sheets. The machine not only adds, subtracts, multiplies, divides, and "compares" quantities, but also, when necessary, it "consults" its "memory" of past operations and refers to stored mathematical tables.

The machine may be arranged to perform a series of steps necessary to obtain a rapid approximation to any logarithm. It can be made to compute various mathematical functions. It can evaluate definite integrals. It can solve differential equations. It can be adapted to solve numerous kinds of problems for mathematicians, physicists, and engineers.

A new striking development in the construction of large multi-purpose digital computers occurred shortly after the appearance of the Harvard Mark I machine. A new computer employing electronic elements for counting and other operations was made at the University of Pennsylvania in Philadelphia. It was christened Electronic Numerical Integrator and Calculator, or ENIAC. It is this machine which began to be referred to in the press as the "electronic brain." In virtue of its electronic operations the ENIAC is extremely fast; this is a great advantage when very long sequences of long calculations have to be performed.

ENIAC was originally used for solving mathematical problems, mainly step-by-step integration, arising in ballistics and aeronautical engineering. Data to be worked upon may be fed into and the results obtained from the machine in several ways, broadly as in the Harvard computers. The machine is not as fast as might be expected because the operation of division is relatively slow (50 operations a second). The arranging of operation sequences, or programming as it is called, is easy, but changing programming is comparatively slow. This is partly because ENIAC, though a multi-purpose machine, was originally designed for a special set of problems.

Since then, that is during the last few years, a considerable number of multi-purpose large electronic digital computers have been or are being built in America, in Britain and elsewhere. The latest machines differ in many ways from the earlier ones, for technical development in this field is rapid.

Mathematical problems in ballistics were some of the first to be tackled with the aid of electronic digital computers. There are now many fields of science and technology in which these machines have been or are likely to be used. Astronomy and Crystallography are two of them in "pure" science. In Astronomy the computation of orbits from sets of observations could and probably will eventually be done by machinery. In Crystallography the determination of the atomic structure of molecules is based upon a mathematical analysis of X-ray spectra, and this is nowadays done much more speedily than ever before with the aid of electronic computers. In Meteorology, weather forecasting could be improved if more elaborate computations upon recorded information were to be carried out. Hitherto such calculations could not have been made in short enough time to be useful. The advent of electronic digital computers has now made the making of these calculations practicable.

In Engineering, extensive calculations are often necessary in arriving at a design which is a compromise between various conflicting technical and economic considerations. It appears that in many cases the designers' work could be speeded up by the use of automatic machinery. In some cases (structural problems, electron-trajectory problems, problems of stability of electrical power networks, and others) this has actually been successfully tried. Automatic electronic digital computers are capable of dealing with statistical problems encountered in government departments; and, as noted earlier, accounting problems of some large industrial and commercial firms have begun to be similarly tackled.

Solving Logic and Game Problems

The interest in the mechanical solution of logical problems is of old standing. Associated with it is the interest in mechanical (and electrical) devices for playing games such as draughts or chess. Such problems can sometimes be tackled by multi-purpose digital machines, but are better handled, or are thought to be capable of being handled, by specially constructed ones. These devices for "mechanical reasoning" can be of relatively simple construction. They are circuits consisting of switching or "on-off" elements such as electro-magnetic relays. "On" and "off" can, of course, stand for *1* and *0*, or yes and no, or right and wrong.

In the late eighteenth century one Lord Stanhope constructed a gadget for solving syllogisms. However, the more recent developments are probably of greater immediate interest. The noted economist of the last century, W. S. Jevons, concerned himself among other things with Boolian symbolic logic (or mathematical logic). Eventually he arranged for a machine to be built—he called it the "Logical Piano"—which was capable of solving mechanically Boolian algebraic equations. Those who are not repelled by phrases such as "mechanical brains" and "electronic brains" might well call Jevons's machine an early "mechanical brain."

Since those days, a number of mechanisms for solving syllogisms and other simple logical problems have been built in Britain, on the Continent of Europe (notably in Italy), and in America. Latterly, electrical circuits have been used in preference to purely mechanical arrangements. In 1938 C. E. Shannon of the Bell Telephone Laboratories, New Jersey, showed how a binary system of number notation and coding can be adapted for the purpose of manipulating the "true-false" relationships of logic.

This work has led to the development of the so-called Kalin-Burkhart Logical Truth Calculator for dealing with symbolic logic problems. In principle it is not dissimilar from Jevons's Logical Piano. Quite independently a "reasoning" mechanism known as the Ferranti Logical Computer, capable of dealing with problems of the same nature as the Logical Truth Calculator, was built in Britain. The Philosophy Department in the University of Manchester has its own small "logic machine." A more elaborate machine was built not long ago in New York. Others have been and are being built.

The more primitive logical computers are of the "scanning" type. A machine of this type deals with a problem containing a number of two-state logical variables by going over all the combinations of the states of the variables in a prearranged sequence. A more "sophisticated" type of computer is one in which this rigid search is avoided. This is achieved in the so-called Feedback Logical Computer. Here the scanning sequence depends upon the particular character of the problem. The superiority of the Feedback Logical Computer is more marked with certain classes of problem than with others. Adaptability or variability of action in response to the nature of the situation may be said to characterize intelligent behaviour; this is of greater advantage in some situations than in others.

Attempts to construct machines to play games go quite a long way back. Babbage, famous for the design of the Analytical Engine, proposed a mechanism to play noughts-and-crosses. During the Festival of Britain a machine that played the quite complex game NIM was exhibited.

Much has recently been said about constructing a chess-playing machine. Wiener, Shannon, Ashby, and many others have been interested in the scheme. Such a machine, although theoretically possible, has not yet been designed. Shannon has actually discussed the programming of a digital computer to play chess. It has been suggested that the discrete nature of chess fits in with the digital character of modern computing machines. Very broadly, a machine to play chess would have to be supplied with criteria of bad and good moves. Each time it was its turn to move, it would have to "look ahead" two or even three moves. It would perhaps go over all the possibilities, and select the one which is best. A machine along these lines could be made to play chess quite as well as would do the most consistent looking ahead for, say, two moves. Unfortunately, the machine would not learn to play better by its mistakes; it would not, that is, profit by experience. Some of those concerned with this field have been expressing the view that it appears as if perhaps an analogue machine (or a combination of digital and analogue machines) would be more likely than a purely digital one to prove a satisfactory robot chess player.

Analogue Machines

In analogue computing machines (or analogy machines) numbers are represented by physical quantities, the magnitudes of which the numbers determine. Mathematical operations are represented by physical events which consist of manipulations of the physical quantities. The result of a set of manipulations is a further physical quantity and the measure of its magnitude stands for the number which is the solution to the problem. In an analogical device, then, discrete numbers are represented by the continuous variation of some physical entity, for example a distance, or an electrical resistance. Analogy machines are characterized by continuous operations in sharp contrast with the discrete operations of the digital computers. And there is a complete analogy between physical quantities and events on the one hand, and numbers and mathematical operations on the other.

A simpler-sounding term for "analogue machines" is "mathematical instruments." However, the use of the word "analogue" or "analogy" in this connection is an apt one; it is universal in America, and it has gained ground in Britain, particularly when applied to large and complex pieces of equipment.

An example of a mathematical instrument is the slide-rule. To multiply one number by another, two distances representing the numbers are placed on end, i.e., added together: this is the continuous operation in this case. Consider, however, another continuous operation used to perform multiplication. In this one, the numbers to be multiplied together are represented by two voltages; the voltages are applied to two sets of input terminals of an electric circuit. The circuit can be so designed that the magnitude of the output voltage corresponds to the product of the two numbers.

Theoretically, any physical quantities may be employed so long as they can be made to obey such laws as will represent adequately the mathematical operations that are required. In practice the commonest media of working are linear displacements, rotations, and electrical resistances, currents, and voltages. Purely electrical methods have the advantage of mechanical simplicity (no moving parts) and high speed of operation. But the scope of calculations may be greatly increased by supplementing electrical arrangements with mechanical ones.

There are analogy computing machines to perform addition, subtraction, multiplication, and division, and also differentiation and integration. There are analogy machines for solving simultaneous linear equations. Trigonometrical and other functions of variables may be obtained. Above all, analogy machines are made to solve differential equations. Altogether, a wide range of problems important in Physics and Engineering can be tackled.

The original conception of such a machine was due to Lord Kelvin in the eighth decade of the last century. But it was not until the nineteen-twenties that the idea was seriously taken up as a practical proposition. V. Bush, of the Massachusetts Institute of Technology, was the first to build such an instrument. Before long there was one built in Manchester and a similar one in Cambridge. In 1945 there was a new machine made at the Massachusetts Institute. The integrating elements of this new machine are mechanical, like those of its predecessor, but its interconnections are now electrical.

The M.I.T. Differential Analyzers and other machines of this type have been used for numerous problems arising, for example, in such studies as those of rotating electrical machinery, non-linear electric circuits, effects of time-lag on the performance of automatic controllers, oscillations of the atmosphere, fluid flow, chemical kinetics, heat conduction, movement of electrically charged particles in magnetic fields, atomic structure calculations, and so forth.

The Homeostat

Late in 1948 there appeared in an engineering journal an article by W. R. Ashby describing a piece of electrical equipment which he had constructed. This apparatus, called the "Homeostat," was later described again in an article in a symposium on Neuropsychiatry, and then in 1952 in a book entitled *Design for a Brain*. The Homeostat, unlike the previously described machines, was specifically designed to imitate certain features—in fact, it is said, the vital features—of the behaviour of an animal, or the behaviour of the nervous system of an animal.

The Homeostat may be classified as an analogue computing machine; it has been said that it is hardly more like the brain than many other machines. Whether or not this is the case, the creation of the Homeostat certainly draws attention in a forceful manner to the fact that some supposedly exclusive features of the behaviour of living things appear in fact to be rooted in a mechanistic principle.

In brief, the Homeostat consists of four units or boxes. On top of each is pivoted a magnetic needle. Each of the four little magnets is situated in the electro-magnetic field of its coil. Each magnet, which can be deflected from its central position, carries a wire which dips in a water potentiometer. All the four units are electrically interconnected: the movements of the magnets modify the coil currents, which in turn modify the magnet movements, and so on.

The method of the interconnections within this arrangement ensures that the magnetic needles are in a state of stable equilibrium. The magnets, when deflected, always return to the central position. This is their fixed end-state or "goal." But, according to circumstances, they reach this goal in different ways, as occasion demands.

The interconnections between the four units (the first order feedbacks) are continuously changed by a multirotating contactor until a position of stability is attained. This Ashby describes as Second Order Feedback. Even partial damage of the apparatus does not stop it from functioning in this way. Ashby calls this property "ultra-stability." He asserts that the principle of ultrastability embodied in his apparatus is of very great general importance.

"Mechanical Animals"

Some thirty years ago an interesting little toy enjoyed a good deal of popularity. It was known as the mechanical "beetle." When wound up, it could move on toothed wheels across a table. The "beetle" was provided with a pair of "feelers," one of which slid along the table top. On reaching the edge of the table, the sliding feeler would drop; this would operate the toy's mechanism in such a way that the beetle would change the direction of its movement to that parallel with the table edge. It would again move on until on reaching the corner, it would change its direction once more, and so on. The mechanism was quite simple. The behaviour of the "beetle" was no more complex than that of many other mechanical devices which respond automatically to certain signals, clues, or stimuli.

It could be maintained that all reflex responses of animals and men to environmental stimuli are as automatic as those of the mechanical "beetle." The point of interest is whether and to what extent relatively complex animal behavior can be imitated by mechanical and electrical artefacts. Grey Walter has in recent years built, exhibited, and written about just such devices.

Grey Walter's "model organism," or "Machina speculatrix," or "tortoise"—the device has been known by all these names—receives signals from its environment through a contact receptor and through a photo-electric cell receptor. An electric motor, controlled by two electro-magnetic relays which are energized by the incoming signals, drives the wheels of the device. Other components mounted upon the chassis include a 6-volt accumulator, two miniature thermionic valves, some resistors and condensers, and a pilot light. The photo-electric cell is made to turn round and round by a small "scanning" motor; the photo-receptor is, thus, in continuous rotation, scanning, as it were, the horizon for light signals. The scanning arrangement controls the device's steering mechanism. The "tortoise" normally wanders around or "explores" the field, but on receiving a light signal it responds by moving towards it.

Despite its structural simplicity, Machina speculatrix exhibits quite a range of behavior generally thought to be characteristic of living creatures. Amusingly enough, it displays what amounts to positive and negative tropism, discriminatory responses, search for environmental optima, or even behavior reminiscent of self-recognition in a mirror or a recognition of another of its kind. The comparative diversity of behavior of the "model animal" is due not to any uniqueness of the component elements but rather to the ingenuity of their interconnection.

Learning to Run Mazes

For over a quarter of a century animal psychology has used rats and sometimes other animals to run mazes. Stylus mazes, which blindfolded human subjects learn to solve with their fingers, are also customarily found in psychological laboratories. Animals and human beings appear to tackle the solution of mazes by trial and error. Can this process be imitated by machine?

In 1938, T. Ross, in America, described a device capable of running and learning a simple maze. The maze runner moved on a network of toy train tracks. Another such device, rather more elaborate, was built in 1952 (also in America) by R. A. Wallace. It, too, ran on rails. Once it solved a maze, it would "remember" the solution and, on a subsequent occasion, run straight through without making errors.

The versatile Shannon, too, interested himself in the construction of a mechanical maze solver. He demonstrated in 1951 a maze-solving machine of somewhat different construction. A panel of 25 squares may be made into a maze by fixing a set of movable partitions upon it. The maze may be altered at will by a rearrangement of the partitions. The maze is explored by a "sensing finger" which can "feel" the walls of the maze as it comes against them. The machine has to guide the finger through the maze to the goal. The goal, in the form of a pin, can also be moved into any of the 25 squares. Shannon's maze solver runs the maze for the first time following the "exploration strategy." Its errors are registered by its "memory." In the second run, the maze solver follows the "goal strategy" and makes no further errors.

In 1950, entirely independently and unaware of any previous such attempts, I. P. Howard of the University of Durham began constructing a model "rat," or an electro-mechanical maze runner, as it has been called; he described it fully in 1953. Howard's "rat" will run any maze provided the width of the lanes is within certain limits. It runs on three wheels and has three spring-loaded "feelers," one on each side and one in front. Its "body" consists essentially of small motors, a set of electro-magnetic relays and a "memory" wheel. Like Shannon's maze solver, the "rat," when placed at the entrance of the maze, begins a systematic exploration; it has been arranged that it should do this by always turning to the left in the first place. The "rat" eliminates all the blind alleys, and in doing so registers its "errors" on the "memory" wheel which controls the settings of the various relays. On a second and subsequent runs, the "rat" makes no further mistakes. Howard's maze runner can learn any maze. The mazes used can be the same as those used by live rats. Such mazes are conveniently rearranged by the shifting of partitions which have legs plugging into holes in a large metal panel.

More recently, J. A. Deutsch of the University of Oxford has constructed a "machine with insight." This, the most advanced of its kind, can not only learn simple mazes, but takes advantage of short-cuts when such are introduced. It can also learn two mazes, and when the two share a common point, the machine can find its way "to whichever of the two goal boxes the demonstrator makes it seek" without any further trial and error.

Our final selection is probably the most striking example of
the application of mathematics in human history. Dated
1905, this paper by Albert Einstein is a statement of the
equivalence of energy and mass. The lay reader will have
difficulty following Einstein's mathematical reasoning. He will
be totally devoid of imagination if he does not see, implicit
in the equations, the shadows they cast before them of
Hiroshima, the thermonuclear bomb, the problem of radio-
active poisoning, and the whole dilemma of modern man.
The translator, Harlow Shapley, is Professor Emeritus of
Astronomy at Harvard, the coeditor of the well-known
Treasury of Science, the recipient of numerous scientific awards,
and one of the most distinguished of contemporary scientists.

THE E = Mc² EQUATION

ALBERT EINSTEIN

THE RESULTS of my recently published electrodynamic
investigation lead to a very interesting conclusion, which is
here to be deduced.

I based that investigation on the Maxwell-Hertz equations
for empty space, together with the Maxwellian expression
for the electromagnetic energy of space, and in addition
the principle that:—

The laws by which the states of physical systems change
do not depend on which of two systems of coordinates, in
uniform motion of parallel translation relatively to each
other, these alterations of state are referred to (principle of
relativity).

With these principles [1] as my basis I deduced *inter alia*
the following result:—

Let a system of plane waves of light, referred to the
system of coordinates $(x, y, z,)$, possess the energy l; let

1.   The principle of the constancy of the velocity of light is of
course contained in Maxwell's equations.

the direction of the ray (the wave-normal) make an angle ϕ with the axis of x of the system. If we introduce a new system of coordinates (ξ, η, ζ) moving in uniform parallel translation with respect to the system $(x, y, z,)$, and its origin of coordinates moving along the axis of x with the velocity v, then this quantity of light—measured in the system (ξ, η, ζ)—possesses the energy

$$l^* = l \, \frac{1 - \dfrac{v}{c} \cos \phi}{\sqrt{1 - v^2/c^2}}$$

where c denotes the velocity of light. We shall make use of this result in what follows.

Let there be a stationary body in the system (x, y, z), and let its energy—referred to the system (x, y, z)—be E_0. Let the energy of the body relative to the system (ξ, η, ζ), moving as above with the velocity v, be H_0.

Let this body send out, in a direction making an angle ϕ with the axis of x, plane waves of light, of energy $\frac{1}{2} L$ (measured relatively to (x, y, z)), and simultaneously an equal quantity of light in the opposite direction. Meanwhile the body remains at rest with respect to the system (x, y, z). The principle of energy must apply to this process, and in fact (by the principle of relativity) with respect to both systems of coordinates. If we call the energy of the body after the emission of light E_1 or H_1 respectively, measured relatively to the system (x, y, z) or (ξ, η, ζ) respectively, then by employing the relation given above we obtain

$$E_0 = E_1 + \tfrac{1}{2} L + \tfrac{1}{2} L,$$

$$H_0 = H_1 + \frac{1}{2} L \, \frac{1 - \dfrac{v}{c} \cos \phi}{\sqrt{1 - v^2/c^2}} + \frac{1}{2} L \, \frac{1 + \dfrac{v}{c} \cos \phi}{\sqrt{1 - v^2/c^2}}$$

$$= H + \frac{L}{\sqrt{1 - v^2/c^2}}.$$

By subtraction we obtain from these equations

$$H_0 - E_0 - (H_1 - E_1) = L \left\{ \frac{1}{\sqrt{1 - v^2/c^2}} - 1 \right\}.$$

The two differences of the form $H - E$ occurring in this expression have simple physical meanings. H and E are energy values of the same body referred to two systems of coordinates which are in motion relatively to each other, the body being at rest in one of the two systems (namely, (x, y, z)). Thus it is clear that the difference $H - E$ can differ from the kinetic energy K of the body, with respect to the other system (ξ, η, ζ), only by an additive constant C, which depends on the choice of the arbitrary additive constants of the energies H and E. Thus we may place

$$H_0 - E_0 = K_0 + C,$$
$$H_1 - E_1 = K_1 + C,$$

since C does not change during the emission of light. So we have

$$K_0 - K_1 = L \left\{ \frac{1}{\sqrt{1 - v^2/c^2}} - 1 \right\}.$$

The kinetic energy of the body with respect to (ξ, η, ζ) diminishes as a result of the emission of light, and the amount of diminution is independent of the properties of the body. Moreover, the difference $K_0 - K_1$, like the kinetic energy of the electron, depends on the velocity.

Neglecting magnitudes of fourth and higher orders we may place

$$K_0 - K_1 = \frac{1}{2} \frac{L}{c^2} v^2.$$

From this equation it directly follows that:—

If a body gives off the energy L in the form of radiation, its mass diminishes by L/c^2. The fact that the energy withdrawn from the body becomes energy of radiation evidently makes no difference so that we are led to the more general conclusion that:

The mass of a body is a measure of its energy-content; if the energy changes by L, the mass changes in the same sense by $L/9 \times 10^{20}$, the energy being measured in ergs, and the mass in grams.

It is not impossible that with bodies whose energy-content is variable to a high degree (*e.g.*, with radium salts) the theory may be successfully put to the test.

Gibt ein Körper die Energie L in Form von Strahlung ab, so verkleinert sich seine Masse um L/V^2. Hierbei ist es offenbar unwesentlich, dass die dem Körper entzogene Energie gerade in Energie der Strahlung übergeht, so dass wir zu der allgemeineren Folgerung geführt werden:

Die Masse eines Körpers ist en Mass für dessen Energieinhalt; ändert sich die Energie um L, so ändert sich die Masse in demselben Sinne um $L/9.10^{20}$, wenn die Energie in Erg und die Masse in Grammen gemessen wird.

Es ist nicht ausgeschlossen, dass bei Körpern, deren Energieinhalt in hohem Masse veränderlich ist (z. B. bei den Radiumsalzen), eine Prüfung der Theorie gelingen wird.

Wenn die Theorie den Tatsachen entspricht, so überträgt die Strahlung Trägheit zwischen den emittierenden und absorbierenden Körpern.

Bern, September 1905.

Einstein's original statement of the mass-energy equivalence.